Prai
The Moon Watche

"Donna Henes takes you on a scenic moonlight drive through time and space, cultures, places and phases: around the world to the moon and back! A delightful collection of all things lunar, encompassing poetry, moonlore, and hard scientific fact. A fine addition to any moon watcher's library."

—Nancy FW Passmore,
editor of *The Lunar Calendar: Dedicated to the Goddess in Her Many Guises*

"Henes' writing is always entertaining. *The Moon Watcher's Companion* is ideal if you love the moon and can't get enough of Her."

—*Beltane Papers*

Also by Donna Henes:

The Queen of My Self:
Women Stepping Into Sovereignty in Midlife

Dressing Our Wounds in Warm Clothes:
Ward's Island Energy Trance Mission

Celestially Auspicious Occasions: Seasons,
Cycles and Celebrations

Reverence to Her. Part I: Mythology, the Matriarchy & Me (CD)

Always in Season: Living in Sync with the Cycles
(Quarterly Journal)

The Moon Watcher's Companion

Everything You Ever Wanted to Know About the Moon, and More

By Donna Henes

Marlowe & Company
New York

THE MOON WATCHER'S COMPANION:
Everything You Ever Wanted to Know About the Moon, and More

Published by
Marlowe & Company
An Imprint of Avalon Publishing Group Incorporated
245 West 17th Street • 11th Floor
New York, NY 10011-5300

Library of Congress Cataloging-in-Publication Data is available.

ISBN 1-56924-466-9

9 8 7 6 5 4 3 2 1

Designed by Brenda Fortunato, Jennifer Collins, and Marina Bekkerman

Printed in the United States of America
Distributed by Publishers Group West

The moon has a face like the clock in the hall;
She shines on thieves on the garden wall,
On streets and fields and harbor quays,
And birdies asleep in the forks of trees.

The squalling cat and the squeaking mouse,
The howling dog by the door of the house,
The bat that lies in bed at noon.
All love to be out by the light of the moon.

But all of the things that belong to the day
Cuddle to sleep to be out of her way;
And flowers and children close their eyes
Till up in the morning the sun shall rise.

Robert Louis Stevenson

Introduction

This book was born in the head of William Galison.

A decade ago, William contacted me after reading several pieces that I had written about the moon for my syndicated column "Celestially Auspicious Occasions."

He told me that he was in the process of inventing a completely revolutionary wrist-watch with a face that would change daily along with the phases of the moon.

The MoonWatch would display the cycles of the moon in order to make them more intimately accessible. No matter where one was, or what the weather, it would be possible to see the moon. Being constantly aware of the moon offers us a connection to the grandeur of the universe.

William felt that knowing something about the science and the mythology, the multi-cultural lore surrounding the moon and its phases would greatly enrich the experience of tracking its changes.

"Would you be interested in writing a companion book for the MoonWatch?" he inquired.

"Would I!" I replied. And the rest is history.

Ten years and an infinite amount of innovative work later, the marvelous MoonWatch has made its debut and *The Moon Watcher's Companion* along with it.

All thanks to William for his insight, foresight, inspiration and support.

May we live by the light of the silvery moon.

Donna Henes
Exotic Brooklyn, NY

Contents

B
A

Faces in the Moon

How It Appears To Us

Moon Watching

Lunar Riddles

England

Q. In Mornigan's park there is a deer,
Silver horns and golden ear,
Neither fish, flesh, feather nor bone,
In Mornigan's park she walks alone.

A. The crescent moon.

Native North America

Q. What has two horns when young,
Loses them in middle age,
And regains them in old age?

A. The moon.

Sweden

Q. Father's sickle is hanging on
Mother's Sunday skirt.

A. The new moon.

Moon Watching

Wondering About the Moon

))))●●●●●((((

FROM THE FIRST OF TIME,
BEFORE THE FIRST OF TIME, BEFORE THE
FIRST MEN TASTED TIME, WE THOUGHT OF YOU.
YOU WERE A WONDER TO US, UNATTAINABLE,
A LONGING PAST THE REACH OF LONGING,
A LIGHT BEYOND OUR LIGHT, OUR LIVES - PERHAPS
A MEANING TO US.

From "Voyage to the Moon"
Archibald MacLeish
20th Century American

In the beginning, there was the moon. Ever present and continually changing. Earth's constant companion on the course of its journey through space, dancing in circles around us like a puppy. Loyal moon; never leaving our side except for those three disconcerting days when it goes about its secret, invisible way, keeping us in the dark.

The moon has always provided humankind with a mysterious and fascinating nocturnal sky show, endlessly interesting to speculate about. From the very earliest time, people have watched and wondered. What is it made of? How was it formed? Why does it behave as it does? What does it look like up close? Who lives there?

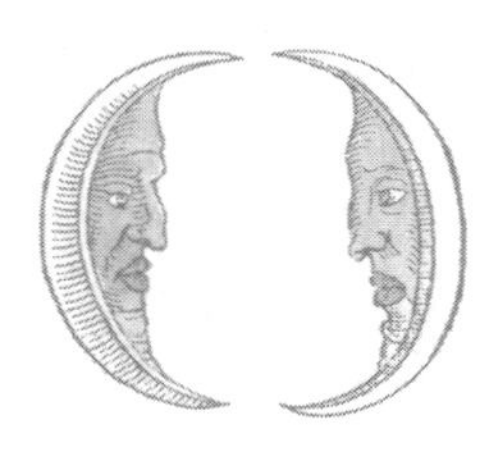

Lunar Inhabitants

The Man in the Moon

The man in the moon is a folkloric motif in much of the world. The Sanskrit name for moon, *mâs*, is masculine, as is *mâni*, the Norse. The moon is male in all Teutonic languages, and although the moon is referred to as female in English, French, Spanish, Italian, Latin and Greek, the image that is commonly conjured is of the guy who hangs out up there.

> THE MAN IN THE MOON CAME TUMBLING DOWN,
> AND ASKED THE WAY TO NORWICH;
> HE WENT BY THE SOUTH, AND BURNT HIS MOUTH
> WITH EATING COLD PEASE AND PORRIDGE.
>
> *Mother Goose*

He is Moon Man in Australia, Malaysia and other parts of the South Pacific. Yue-Lao is the Chinese Old Man in the Moon. He is the celestial matchmaker who predestines earthly marriages and binds together couples-to-be with lengths of red silk cord. Jewish Talmudic tradition holds that Jacob is in the moon, while French superstition tells that it is Judas Iscariot. In ancient Egypt, he was Thoth, the dog-faced ape.

> PRAISE TO THOTH, THE SON OF RA,
> THE MOON BEAUTIFUL
> IN HIS RISING, LORD OF BRIGHT APPEARINGS
> WHO ILLUMINATES THE GODS.
>
> *Egyptian Hymn*

The Serbian moon is called Myesyts in his guise as the Bald Uncle, but s/he is also referred to in the feminine familiar as the Pretty Little One, an expression of identity ambivalence. The Chaga of Kenya see the paleness of lunar light as evidence that the Moon Chief and his below-par followers are too backward to know how to use fire.

Roong is the male deity of the moon to the Haida of western Canada. In his loneliness, he abducts a man to join him in the night sky. When, every so often, the captured man attempts to escape, he spills his pail of water which rains down on the earth.

SCOOPING UP THE MOON
IN THE WASH-BASIN,
AND SPILLING IT.

Ryoho
18th Century Japanese

The Norwegian *Edda* describes the visible spots on the moon as a boy and a girl carrying a bucket of water. The children, Hjuki and Bil, have been adopted into our lexicon of legend as Jack and Jill. Their trudge up and consequent trip down the hill traces the monthly ascent and descent of the phases of the moon.

JACK AND JILL
WENT UP THE HILL
TO FETCH A PAIL OF WATER.
JACK FELL DOWN AND
BROKE HIS CROWN AND
JILL CAME TUMBLING AFTER.

Mother Goose

Many peoples blame the periodic changes in the vestige of the moon on its bad behavior, unbecoming to an officer or a gentleman. In Germany, the Man in the Moon carries a bundle of firewood, cabbages, sheep or straw, that he has stolen. He is a thief and worse. In Europe of old, people thought that the dirty old Man in the Moon, with his parts fully engorged, impregnated women, making it unsafe to sleep in the moonlight. In a complete role reversal, Siberian Chuckchi men exhibit their privates to the female moon when they pray for her power.

Uaupe Indians of the upper Amazon believe that a young woman's first menstruation is the result of her being de-flowered by the moon. Similarly, in Papua New Guinea they say that the moon is a girl's first husband and her monthly blood is proof of their relations.

According to the Khasias tribes of the Himalayan mountains, the moon keeps falling in inappropriate love with the mother of his wife, the sun. His monthly darkening is attributed to the ashes which his indignant mother-in-law hurls in his face in response to his offensive indiscretions.

An equivalent tale is told among the Eskimos who inhabit the northernmost reaches of North America, Greenland and Siberia. Here, the sun and moon, Malina and Anninga, are sister and brother. One night, Anninga secretes himself into his sister's chamber and surprises her in an impure embrace. Shocked, but undaunted, she rubs soot on his face in the dark so that she might recognize her attacker by the light of day. The moon is thus found out. And so, rebuked, he must forever remain in the shadow of his sister's superior light.

Versions of this story are told throughout the Americas as far south as Brazil. In Romania, the roles are reversed. In this case, the moon is female. She dirties her own face with ashes to make herself unappealing in an attempt to repel the unwanted incestuous advances of her brother, the sun. According to Slavonic legend, the moon, the King of the Night, was unfaithful to the sun who punishes him periodically for his transgressions.

The Moon Man of the African Bushmen, too, incurs the wrath of the sun who slices away at him, sliver by sliver, with a sword until he is completely destroyed. Exhausted after his ordeal, he has to go away to rest and regain his strength so that he may reappear the following month.

Lunar Inhabitants

Moon Mother

The moon as mother is a prevalent mythological theme, more primal than, and, in many cases, predating that of the marauding male moon.

The first woman of Polynesia was the moon, Hina, and each woman thereafter is a *wahine*, created in her image.

> HINA-WHO-WORKED-IN-THE-MOON
> FLOATED AS A BAILER. SHE WAS TAKEN INTO A
> CANOE AND CALLED HINA-THE-BAILER.
> SHE WAS CARRIED TO THE SHORE AND PUT BY THE FIRE.
> CORAL INSECTS WERE BORN, THE EEL WAS BORN,
> THE SEA-URCHIN WAS BORN,
> THE VOLCANIC STONE WAS BORN.
> SO SHE WAS CALLED
> HINA-FROM-WHOSE-WOMB-CAME-VARIOUS-FORMS.

Cumulipo Tribe, Oceanic Isles

Ix Chel, the Mayan Moon Goddess, was the first woman of the world and the mother of all deities. She is the giver of the life-enhancing waters, and the protector of women in childbirth. The Moon Goddess of the Chuckchi Siberians is called Mother, and their Eskimo relatives call her who shines so brightly, She Who Will Not Take a Husband.

> OH MOON, OH MOON! WHO IS YOUR MOTHER?
> WHITE CRESCENT! IF SHE FROWNS AT YOU – BAD HARVEST
> BEFALLS YOU. IF SHE DOES NOT – YOUR MOUTH WILL BE
> FULL. SHE GAVE ME CHARMS FOR WOMAN OF ANY AGE.
> SHE GAVE ME THESE TWO SECRETS – BUT DON'T ASK WHY.

Elema Tribe, Papua New Guinea

Egyptians called the moon the Mother of the Universe, because the moon has, according to Plutarch, "the light which makes moist and pregnant, is promotive of the generation of living beings and the fructification of plants." The Egyptian hieroglyph, *mena*, means both moon and breast. Hathor, the Fertile Sky Goddess, is the Celestial Cow and is depicted as carrying the sun (her son) disk between her horns. From her breasts flow the stars and milky way and all of the waters of life.

Britain was originally called Albion after the Milk-White Moon Goddess. The European continent is named after the goddess Europa, who was also known as Hera and Io, the White Moon Cow. The Finnish creatrix was known as Luonnotar, Luna the Moon. It was she who gave forth the great World Egg from which hatched the entire universe.

The Peruvian moon was Mama Quilla. She, too, bore an egg. Mama Ogllo, the Moon Maiden, along with her brother, the sun, founded the royal Incan dynasty. The Zuni of the American Southwest venerate the Moon Our Mother, who is the younger sister of the sun. The Sioux call her the Old Woman Who Never Dies. To the Iroquois, she is the Eternal One, the Mother Who Created the Earth and the Surface People. To the Apache and the Navaho, she is Changing Woman.

THE FIRST WOMAN HOLDS IT IN HER HANDS.
SHE HOLDS THE MOON IN HER HANDS.
IN THE CENTER OF THE SKY, SHE HOLDS IT IN HER HANDS.
AS SHE HOLDS IT IN HER HANDS, IT STARTS UPWARD.

THE FIRST WOMAN HOLDS IT IN HER HANDS.
IN THE CENTER OF THE SKY, SHE HOLDS IT IN HER HANDS.
SHE HOLDS THE MOON IN HER HANDS.
AS SHE HOLDS IT IN HER HANDS, IT STARTS DOWNWARD.

Navaho

The moon, as Queen of Heaven, reigned in the Near East: Babylonia, Persia, Syria, Sumeria, Akkadia and Canaan. She From Whom All Life Issues was known as Anath, Asherath, Anahita, Qadesh, Lilith, Ishtar, Inanna and Astarte, which means womb. As Ishtar, she sings, "I, the mother, have begotten my people, and like the young of the fishes, they fill the seas." In Persia, her name was Metra, She Whose Love Penetrated Everywhere.

In pre-Islamic Arabia, the moon was feminine and her cult prevailed. She was Manat, the Moon Mother of Mecca, and her shrines are still holy, although women are now constrained from entering them. Another of her names, Al-Lat, was ultimately altered to become Allah.

The Greek Hera, Demeter, Artemis, Thetis, Phoebe and Selene; the Roman Luna, Mana and Diana; and Gala or Galata of the Gaelic and Gaulish tribes, were also associated with the moon. The Virgin Mary, who is referred to as Queen of Heaven, is frequently displayed standing upon a crescent moon. In central Asia, the moon is said to be the mirror of the Great Goddess which reflects everything in the universe.

GODDESS OF THE PERFECTION OF WISDOM,
HOLY TARA WHO DELIGHTS THE HEART.
FRIEND OF THE DRUM, PERFECT QUEEN OF
SACRED LORE WHO SPEAKS KINDLY.
WITH A FACE LIKE THE MOON, SHINING
BRILLIANTLY, UNCONQUERED. . .

Aryatârâbhaṭṭârikânâmâshṭottarasatakastotra
Hindu Text

The Moon Goddess is the Divine Midwife. The West African Niger believe that the Great Moon Mother sends the Moon Bird to Earth to deliver babies. The Baganda of Central Africa bathe their newborns by the light of the first full moon following birth. In Ashanti tradition, the moon Akua'ba is a fertility figure who aids conception and guarantees sturdy children.

Moon Spirits

ALL HUMANS SHALL DIE -

ONLY THE MOON SHALL BE REBORN.

Nandi Tribe, Uganda

If the moon was seen as the producer and provider of life, she was also widely perceived to be the guide and guardian of death. Life and death are like the opposite phases of the moon, light and dark. Inanna, Sumerian Goddess of the Moon, was the ruler of life and fertility as well as death and rebirth. The Tartars of Central Asia called the moon Macha Alla, Queen of Life and Death.

The Trobriand Islanders said that the female sorcerers associated with the moon were eaters of the dead. According to the Indian *Vedas*, the souls of the dead go to the moon to be consumed by the lunar maternal spirits. The crescent moon is often seen as a boat which ferries the souls of the newly dead to their resting place on the moon.

THE MOON AND THE YEAR

TRAVEL AND PASS AWAY:

ALSO THE DAY, ALSO THE WINE.

ALSO THE FLESH PASSES AWAY

TO THE PLACE OF ITS QUIETNESS.

Mayan

One of the names of the old Mexican Moon Goddess was Lady of the Place of the Dead. The Egyptians thought that heaven was on the moon. The Greek home of the dead, Elysian Fields, was located there. Plutarch commented that "The moon absorbs the souls of the dead, as the earth absorbs the bodies."

Owl Woman's Death Song

In the great night my heart will go out,
Toward me the darkness comes rattling.
In the great night my heart will go out.

Papago

The Romany Gypsies, too, believe that the dead reside on the moon. The kings of Burundi trace their ancestry to the moon deity and believe that they will return to the moon when they die. Polynesians see the moon as the final resting place of dead kings and chiefs, the Guacurus of South America say dead medicine men stay there. For the South American Salivas, the moon is paradise because there are no mosquitoes.

Song of Two Ghosts

My friend
This is a wide world
We're traveling over
Walking in the moonlight

Omaha

Lunar Inhabitants

ABC Deities

The Gods

Chandras	*Hindu*
Indrus	*Hindu*
Japara	*Australian Aborigine*
Khons	*Egyptian*
Klehanoai	*Navaho*
Kuu	*Finno-Ugric*
Luan	*Irish*
Ma	*Persian*
Moon Brother	*Eskimo*
Moon Old Man	*Taos*
Myesyts	*Serbian*
Roong	*Haida*
Sin	*Assyrian, Babylonian, Sumerian*
Soma	*Vedic Hindu*
Tecciztecatl	*Aztec*
Thoth	*Egyptian*
Tsuki-Yomi	*Japanese Shinto*

The Goddesses

A	*Chaldean*
Akua'ba	*Ashanti*
Albion	*British*

Al-Lat	*Pre-Islamic Arabian*
Al-Mah	*Persian*
Anath	*Ugaric*
Annit	*Northern Babylonian*
Aphrodite	*Greek*
Ardvi Sura Anahita	*Persian*
Arianrod	*Welsh*
Arma	*Hittite*
Artemis	*Greek, Amazonian*
Artimpassa	*Sythian*
Asherath	*Semitic*
Astarte	*Semitic*
Auchimalgen	*Chilean Araucanian*
Bast	*Egyptian*
Brigit	*Celtic*
Britomartis	*Cretan*
Brizo	*Aegean, Delos*
Candi	*Hindu*
Caotlicue	*Aztec*
Changing Woman	*Apache, Navaho*
Ch'ang O	*Chinese*
Coyolxanliqui	*Aztec*
Dae-Soon	*Korean*
Diana	*Roman*
Diktynna	*Cretan*
Doris	*Dorian*
Europa	*Cretan*
Gala	*Gaelic*
Gnatoo	*Friendly Island Polynesian*

Gwaten	*Japanese Buddhist*
Hanwl	*Oglala Sioux*
Hard Beings Woman	*Hopi*
Hecate	*Greek*
Hina	*Polynesian*
Hina-hanaia-i-ka-malama	*Hawaiian*
Huitaca	*Columbian Chibcha*
Hun-Ahpu-Ahpu-Mtye	*Guatemalan*
Inanna	*Sumerian*
Ishtar	*Akkadian*
Isis	*Egyptian*
Ix Chel	*Mayan*
Ix-Huyne	*Mayan*
Juno Lucetia	*Roman*
Lalal	*Etruscan*
Leader of Women	*Mohawk*
Lilith	*Sumerian*
Luna	*Roman*
Macha Alla	*Central Asian*
Mah	*Persian*
Mana	*Roman*
Manat	*Pre-Islamic Arabian*
Mama Quilla	*Peruvian Quichen*
Mardoll	*Scandinavian*
Mary, Queen of Heaven	*Christian*
Mawu	*Dahomey Fon*
Mensa	*Roman*
Meztli	*Aztec*

MITI	*Chukchi*
MONA	*Teutonic*
MOONLIGHT-GIVING MOTHER	*Zuni*
MWEZI	*Tutsi*
NUAH	*Babylonian*
NYADEANG	*Nuer*
OLD WOMAN WHO NEVER DIES	*Sioux*
O SHION	*Kalderash Gypsy*
PANDIA	*Greek*
PE	*Pygmy*
PERSE	*Early Greek*
PHERAIA	*Thessalian*
PHEOBE	*Greek*
QADESH	*Canaanite-Syrian*
RABIE	*Indonesian*
RI	*Phoenician*
SARDARNUNA	*Sumerian, Chaldean*
SELENE	*Greek*
SIRDU	*Sirrida*
TAPA	*Polynesian*
THE ETERNAL ONE	*Iroquois*
TITANIA	*Roman*
TSU-YOMI	*Japanese Shinto*
URSULA	*Slavic*
WHITE SHELL WOMAN	*Navaho*
YEMANJA	*Brazilian Mancumba*
YOLKAI ESTAN	*Navaho*
ZIRNA	*Etruscan*

Lunar Bestiary

Along with all the men, maids, mothers, goddesses and gods, the moon is also mythically populated with a wide variety of animals. Creatures are usually associated with the moon by virtue of having some physical or behavioral resemblance, such as those who are nocturnal and living in the lunar light; those who are aquatic, the tides being affected by the moon; those having crescent-shaped horns, reminiscent of the moon's shape in its waxing and waning stages; and those who change their shape like the moon does.

Moon that is a cow, being horned like her,
Moon that is a panther, rapacious of light,
Moon that is a she-bear, a lioness,
Three-headed hound of the moon,
Moon-muse, mother, fountain that rises and falls,
Your daughters do not forget you.
You make their weather. Their blood
Ebbs and flows like the tides you make.

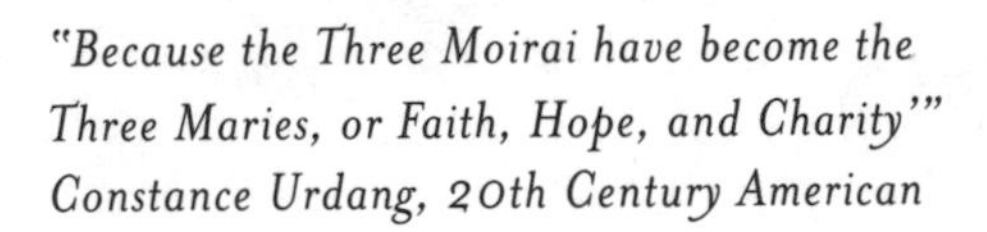

"Because the Three Moirai have become the Three Maries, or Faith, Hope, and Charity'"
Constance Urdang, 20th Century American

Bats

An Alur tale recounts how the King of the Bats invited the Moon for a feast. He offered her succulent meat on a tray, but when the Moon reached for it, he withdrew his hospitality telling her to ask anything else of him but that. The Moon took the tray of meat anyway, for bats are not strong. To this day, bats hang upside with their rear ends pointed to the moon to show their contempt. The origin of mooning, perhaps?

Birds

Certain birds are identified with the moon by virtue of their color or shape or their being aquatic or nocturnal. Thoth, the Egyptian Moon God, was often depicted as an ibis, a water bird. Huitaca, the Moon Goddess of the Cibcha people of Columbia, comes to Earth as an owl.

Birds are also magical because they fly between Earth and sky connecting us with the lunar realm. The West African Niger believe that the Great Moon Mother sends the Moon Bird to Earth to deliver babies.

In East Africa, it is thought that the moon was once much closer to the earth and she looked like a beautiful white bird. When a man shot poisoned arrows at her, she began to wane and soon died.

Bovines

HEY, DIDDLE DIDDLE.
THE CAT AND THE FIDDLE.
THE COW JUMPED OVER THE MOON.
THE LITTLE DOG LAUGHED TO SEE SUCH A SPORT,
AND THE DISH RAN AWAY WITH THE SPOON.

Mother Goose

From India to Scandinavia and in the Western Hemisphere as well, some incarnation of a white Moon Cow Goddess reigned. The eldest of the beneficent bovine sisterhood is Astarte, whom Milton describes as, "Astarte, Queen of Heaven, with crescent horns." She dates from the Neolithic period of the Middle East and continued to flourish in different guises throughout the

Bronze Age. The Egyptian Hathor, the Sky Goddess, is the Celestial Cow and is depicted as carrying the heavenly disk between her horns.

The Canaanite-Syrian goddess, Anath assumed the shape of a cow, as did the Argive Cow-Eyed Hera, the Greek Europa, the Minoan Pasiphae, the Ionian Io the Horned One, the Indian Gaus and the Scandinavian Audumla, all of whom were associated with the moon. Leah, the Hebrew name of the biblical matriarch, mother of the generations, means "wild cow." The Assyrians associated the moon with a cow in their fertility rites. A Sumerian hymn refers to the moon as an "alert bull with untiring hooves."

THE MOON
IS A WHITE
BULL
DELICIOUSLY
STEPPING
ON VELVET

"Images"
Alastair Campbell
20th Century New Zealander

In Namibia, worshippers blow through an antelope horn sounding a greeting to the moon who helps them to hunt nocturnal creatures in the desert. The Roman Moon Goddess was Diana, the Radiant One, who was identified with the Greek Artemis, Queen of the Hunt. Deer and stags were her special creatures. Deer are considered miraculous, because they can regenerate their horns in the way that the moon does.

Cats

Nocturnal, cats are able to see in the dark. This capacity is traditionally related to the concept of lunar lucidity, or inner vision. The black cat represents the dark moon. The sphinx, which stands for the knowledge of the mysteries of life and death, is part lion. The tiger is symbolic of a Chinese monster of the dark night sky.

LOUD AS BACKYARD CATS THE MOON
KNITS HER HAIR OVER HER EYES
WITH MOURNING
POURS WATER ON HER FEET
BRAYS
CHANGES COLOR & MONTH
GIVES UP COFFEE & LATE NIGHT
HOURS
TAKES A LOVER
LOVES HER

THE CAT WALKS THE LEATHER FENCE
OF ITS DREAMS SINGING SONGS
ITS TONGUE IS SWEET
AS GRAPE JELLY AND ITS HOOVES
ARE WET AND SILVER. IT SINGS. AND THE SKY
STANDS ON ONE FOOT, LISTENING. AND THE MOON LICKS
HER FINGERS AND THE CAT
LICKS ITS TOES. IT SINGS. AND THE BRIGHT GREEN STARS
BREATHE THROUGH ITS WHISKERS.
AND THE SNAKE-ROUND MOON SQUATS
ON THE PORCH RAILING CLOSE
TO ITS TAIL. THE CAT SINGS
ITS DREAMS, ITS TURQUOISE AND LAPIS
DREAMS.

Faye Kicknosway
20th Century American

The Egyptians identified the moon with Bast, or Bastet, the Cat Goddess. Cats were sacred to Isis, the Mother Goddess, as well. There is a wide association of the moon with cats throughout East Africa where they — especially sleek black cats — symbolize pleasure-loving young girls.

To the Australian aborigines, the moon, Mityan, was a cat. He fell in love with his neighbor's wife and tried to run away with her, but was caught by the irate husband and beaten. Mityan ran away and they say he has been wandering ever since.

Dogs, Wolves

The dog was the sacred companion of the moon goddess of the prehistoric Balkan civilizations of Old Europe. The early Egyptian moon god, Thoth, was pictured as a dog-faced ape who was slowly eaten away by monsters as he made nightly voyages on his celestial barque. Hecate, the Death Goddess, was associated with the dark moon. She always traveled with dogs. Dogs also accompanied the moon goddess Artemis/Diana, on her nocturnal hunts.

Because they eat carrion, wolves symbolize death and regeneration, the very qualities visible in the cycles of the changeable moon. The Norwegian *Edda* tells of the Goddess Hel who ruled the underworld. Myth has it that she gave birth to lunar wolf-dogs who ate the flesh of the dead before they carried their souls off to paradise.

The Native American Creek believe the moon to be inhabited by an old man and his canine companion. The people of the Seneca Nation say that the Wolf Spirit sang the moon into the sky, and that is why all wolves howl at the moon to this day.

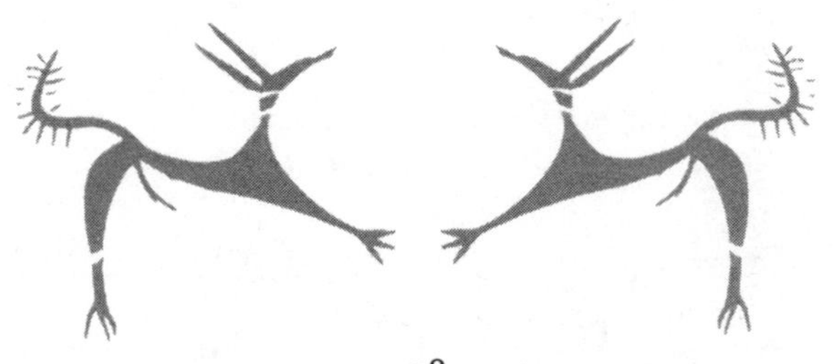

Fish

Fish were traditionally sacred to the Goddess who ruled the moon and the waters. The astrological sign Pisces represents two crescent moons, one waxing, the other waning.

Frogs, Toads

Frogs represent transformation as they are able to change in form, from a tadpole to a frog, and in function, as they can breathe in both air and water. Among Native Americans, the frog is the most commonly recognized image visible in the dark spots of the moon. One legend tells of the frog sisters who ruthlessly reject all of their animal suitors. The forsaken lovers cry a torrent of tears which the frog sisters manage to escape by climbing onto the face of the moon and staying there.

Other stories abound. The frog is the guardian of the sun and the moon which he protects from the bear who might swallow them. The frog swallowed the moon and then the moon turned around and swallowed the frog. If you look at the moon you can see the frog sitting in the center weaving a basket.

The Salish and Maidu of the U.S. Northwest coast see a toad in the moon. They tell the story of how a wolf fell madly in love with a toad. He prayed to the moon for bright light to help him pursue her. Just as he was about to grab the little toad, she made one last chance leap up and landed on the moon, where she stays to this day.

The Chinese story about the toad in the moon goes like this:

> *In the year 2346 B.C., the goddess, Chin Mu, bestowed a pill of immortality upon Shen I, the Divine Archer, in service to the emperor as payment for a house he built for her. She warned him not to take the pill until he had completed*

twelve months of cleansing and meditation in preparation of this great gift.

Shen I hid the pill in the rafters and began his program, but was soon called to battle. In his absence, his wife Ch'ang O noticed a silver white light and sweet smell emanating from the ceiling. The light led her to the pill which she swallowed.

Immediately, she began to rise and float — right out of the window. At this moment, Sheng I showed up. Angry, he chased her with a bow. Her body shrank until she became the size of a toad then she flew up to the moon where she decided to stay.

Even today, votive offerings of toads made of wax, wood and silver are offered in churches in Bavaria, Hungary, Moravia and the former Yugoslavia. They represent the life-giving energy of the moon and the fertile power of the ancient mother goddess.

Insects and Arachnids

The scarab, or dung beetle, was sacred in ancient Egypt. It was believed that although the scarab represented the male sun god, Khepera, it received its power from the moon when it buried its dung ball for 28 days, or a complete lunar cycle. It, like the bee, has crescent-shaped wings.

According to the African Bushmen, the moon was made by a praying mantis from an old shoe. In Central Africa, the moon is associated with the spider. In one Angolan myth, the Moon Princess descended to Earth to marry the Prince of the Land by means of a silver cord woven by the Moon Spider.

Why There are Fireflies

Once upon a time in Japan, there lived a woodsman and his wife who had a pleasant house and beautiful garden at the foot of Mt. Fujiama. The only thing that marred their happiness was the fact that they had no children. The wife prayed to the holy mountain for a child.

Soon a spark of light appeared high up on the mountain and floated down toward them. It was a tiny, perfect Moon Baby sent by the Lady in the Moon. She grew up into a happy, beautiful Moon Princess, adored by her mortal parents and all who knew her. The son of the emperor asked for her hand in marriage, but she refused, explaining that she was bound to return to the Moon Lady, her true mother, after twenty years on Earth.

On the appointed night, the Lady in the Moon sent down a shimmering beam and the princess rode up, weeping all the way for those she left behind. As her silver tears fell, they took wing and drifted all over the land. Her tears can still be seen on moonlit nights as fireflies flit about the sky.

Butterflies and moths are symbols of lunar-like transformation and rebirth as they change from caterpillars. Moths fly around at night searching for the light, ancient symbols for the soul seeking luminous enlightenment.

BUTTERFLIES CREATING FLOWERS
FLOWERS CREATING WATERFALLS
WATERFALLS CREATING RAINBOWS
RAINBOWS CREATING SUNSETS
SUNSETS CREATING MOONLIGHT
MOONLIGHT CREATING BUTTERFLIES.

Kellie Kress, 5th grade
20th Century American

Rabbits

The hare is connected to the moon by the timing of her extremely fecund fertility cycle — roughly that of one moon cycle. Of all the animals in the lunar zoo, the rabbit is the most widespread, recognized in Tibet, Africa, Mexico and throughout the East. The Hindu moon god was known interchangeably as Soma, Indus and Chandras. Chandras is most often shown carrying a hare. The association of the rabbit with the moon traveled from India to China and Japan with the eastward spread of Buddhism.

In China, it is thought that there is a hare on the moon who sits at the foot of a cassia tree grinding the powder of immortality in his mortar. When Ch'ang O, the moon goddess, swallowed this potion-tablet, she ascended to the moon and was transformed into a toad. In Chinese art, the moon is depicted as a white crescent along with a toad and a hare. The Japanese visualize a rabbit pounding rice with a pestle. The ideogram which represents the moon, *mochi-zuke*, and that for the milling of rice into flour for cakes, is the same.

The Hottentots of southern Africa have a myth in which the moon beats a mischievous rabbit with a stick, giving it a hare lip. The native people of old California called the moon the Great Rabbit. Julius Caesar relates that the ancient Britons considered the eating of rabbit taboo due to its lunar relationship. And, as recently as the 1970's, Swabians forbade their children from making hand-shadows on the wall in the image of a bunny for fear of offending the moon. Rabbits' feet are still carried in pockets and purses, backpacks and glove compartments, and hung from rear-view mirrors in sympathetic, if unconscious, moon magic memory.

Most pre-Columbian Mexican codices picture the moon as a crescent-shaped water vessel in which the outlined profile of a rabbit can be seen. The Tezcucan of ancient Mexico said that once upon a time, the sun and the moon were of equal brightness which displeased the gods. So one of them threw a hare at the moon giving it dark bruises on its face in the shape of a rabbit, and forever dimming its light. The lunar hare still prevails in the popular Mexican imagination.

Snakes, Serpents

THE SNAKE SHEDS ITS SKIN, JUST AS THE MOON SHEDS ITS SHADOW. THE SERPENT SHEDS ITS SKIN TO BE BORN AGAIN, AS THE MOON ITS SHADOW TO BE BORN AGAIN. THEY ARE EQUIVALENT SYMBOLS.

Joseph Campbell
20th Century American

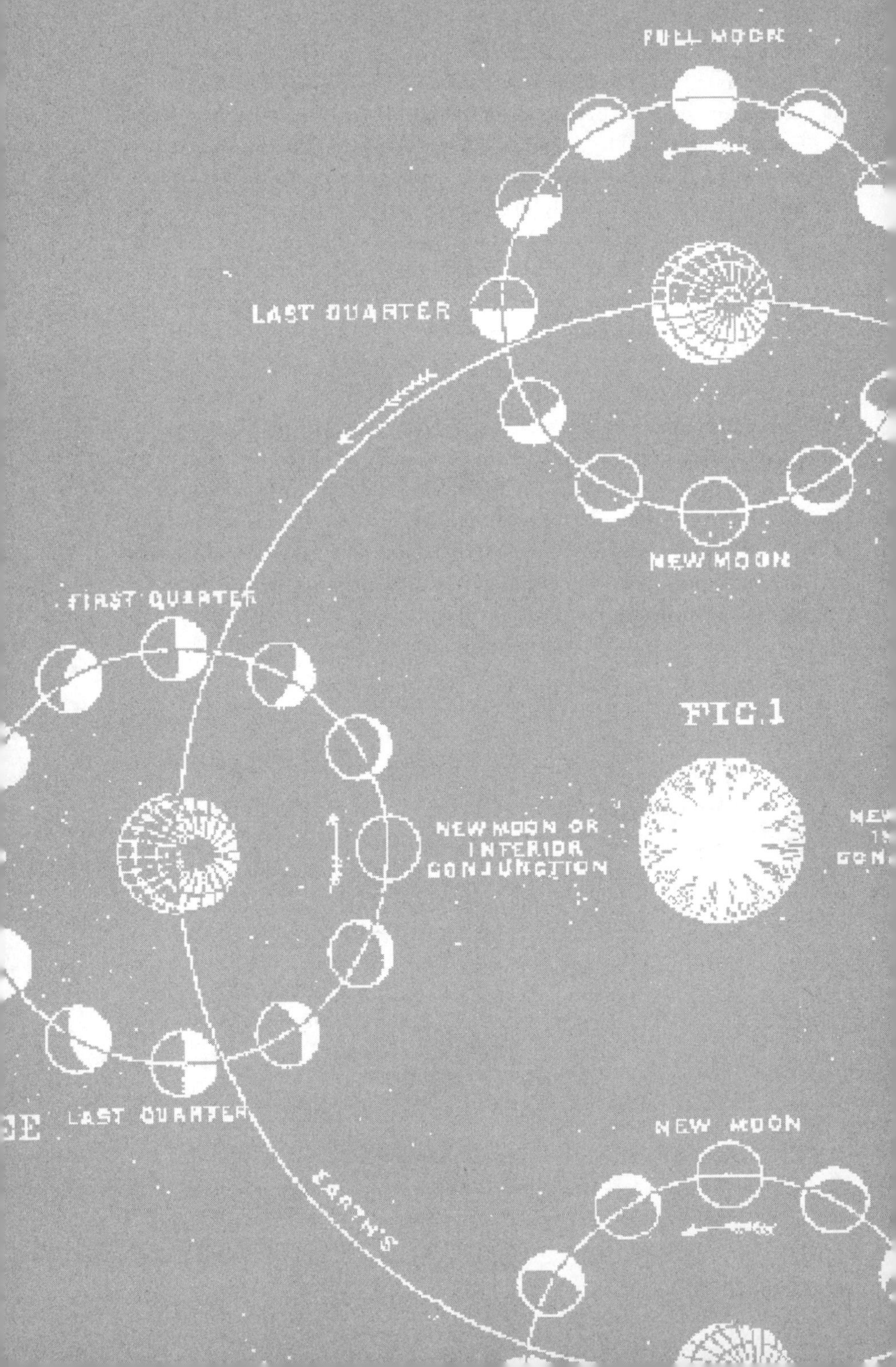

FULL MOON

LAST QUARTER

NEW MOON

FIRST QUARTER

FIG.1

NEW MOON OR
INFERIOR
CONJUNCTION

LAST QUARTER

NEW MOON

EARTH'S

Phases of the Moon

How It Affects Us

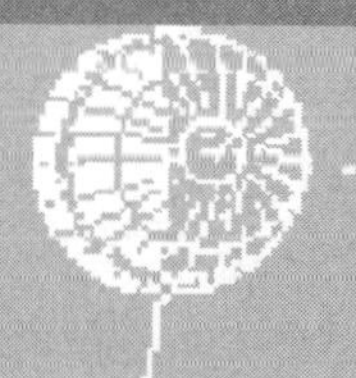

Moon Phases

The Complete Cycle

The great mystery of the moon is its cycle of change, its periodic transformation of size, shape, location. It appears and grows steadily in the dark sky, the effulgence from its light illuminates and animates the night. Then it dims and grows dark, darker, darkest.

It seems to open its silver incandescent eye and then blink it shut again. Again and again, moonth after moonth. It is here and then it is gone. For days at a time, it lies dormant as if dead, invisible to the world, only to return again. And forever and always again, fulfillment, decrease and disappearance.

> Art thou pale for weariness
> Of climbing heaven and gazing on the earth,
> Wandering companionless
> Among the stars that have a different birth,
> And ever-changing like a joyless eye
> That finds no object worth its constancy?
>
> *"To the Moon"*
> *Percy Bysshe Shelley*
> *19th Century English*

The moon is almost always associated with women and is widely considered to be in their special domain. This is due, no doubt, to the clear co-incidence of the periodic cycle of the moon and the monthly hormonal cycle of women. Undeniably, they are all but identical. (Interestingly, very few animals menstruate: only primates, bats, elephant shrews and human beings.) There is a traditional Chinese saying, "Men do not worship the moon, women do not sacrifice to the Kitchen Gods."

The elemental drama of the process of conception through completion, birth through rebirth, is played out within the moon's cycle of perpetual return. Light-dark-light. Egg-blood-egg. Life-death-life. It's born. It lives. It thrives. It dies. It is revived. Its promise of life ever-lasting, fulfilled.

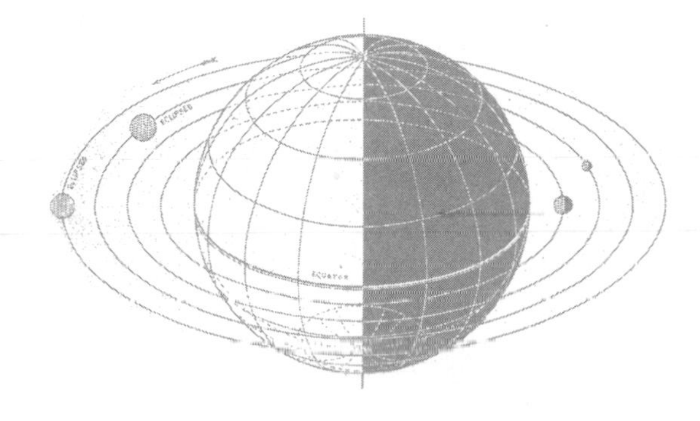

THE HALF MOON SHOWS A FACE OF PLAINTIVE SWEETNESS
READY AND POISED TO WAX OR WANE;
A FIRE OF PALE DESIRE IN INCOMPLETENESS,
TENDING TO PLEASURE OR TO PAIN:—
LO, WHILE WE GAZE SHE ROLLETH ON THE FLEETNESS
TO PERFECT LOSS OR PERFECT GAIN.
HALF BITTERNESS WE KNOW, WE KNOW HALF SWEETNESS;
THIS WORLD IS ON WAX, ON WANE:
WHEN SHALL COMPLETENESS ROUND TIME'S INCOMPLETENESS
FULFILLING JOY, FULFILLING PAIN?-
LO, WHILE WE ASK, LIFE ROLLETH ON IN FLEETNESS
TO FINISHED LOSS OR FINISHED GAIN.

"The Half Moon Shows a
Face of Plaintive Sweetness"
Christina Rosetti
19th Century English

Moon Phases

Waxing Moon

))))●●●●((((

Return soul of the sky, candid moon
To the first sphere, shining and beautiful,
And with your customary brilliance restore
The crown of silver to the darkened sky.

Chiara Cantarini Matraini
16th Century Italian

The return of the first sterling crescent as it rises above the eastern horizon signals the coming to life, the resurrection of the moon, and as such is a time of great optimism and hope. The rising of the new waxing moon inaugurates a new cycle. It sets the pace, the tone, for the new moonth ahead. The crescent moon symbolizes the courage, creativity and determination to begin once again from the beginning.

Crescere means "to grow," related to the Latin, *creare*, "to create," "to produce." Diana was the goddess of the crescent moon to the Romans and the Druids. In her honor, women of Gaul baked crescent-shaped communion cakes to greet the rising of the new moon after three nights of dark skies. These are the same French croissants which enjoy such gustatory popularity today.

When people of the Gold Coast of Africa see the new moon, they throw ashes at it and say, "I saw you before you saw me." They believe that whatever you do on the new moon, you do all moonth long. The men of the Fetu tribe jump three times in the air, clap their hands in rhythms of delight, and raise praise and much thanks.

The first sighting of the new moon still ushers in the new moonth throughout the Middle and Far East with a joyous, thankful celebration for its return to the heavens. *Rosh Chodesh* is the Jewish celebration of the new moon. According to the Jewish Kabbalist tradition, blessings upon the moon are not to be intoned until seven full days after her birth, or the first quarter moon. The supplicant is instructed to glance once at the moon before beginning the sanctifying ceremony, and once the prayer has begun, to refrain from looking again at her face.

A SLIVER OF MOON

...YET A MAIDEN...

IN THE NIGHT SKY

from "WALK in the night"
Karen Ethelsdattar
20th Century American

The Altaic people of Siberia salute each new moon and ask it for happiness and luck. The Grebo of Sierra Leone feed their public talismans and idols rice and oil every new moon in exchange for which they pray that the moon will bring forth abundance, good health and peace.

The Bushmen of Angola beseech the new moon for rain, game for the hunters and an abundance of fruit for gathering, none of which is possible without the moon's protection. The Chagga of the Mt. Kilimanjaro region and the Kundu in Cameroon pray at the new moon for renewal and health. The Kalderash Gypsies intone to the new moon:

THE NEW MOON HAS COME OUT.
MAY SHE BE LUCKY FOR US.
SHE HAS FOUND US PENNILESS.
SHE MAY LEAVE US WITH GOOD FORTUNE
AND WITH GOOD HEALTH, AND MORE.

Moon Phases

FULL MOON

WHEN THE MOON IS FULL IT IS AS IF THE ETERNAL
LIGHT OF THE GREAT SPIRIT WERE UPON THE WHOLE WORLD.

Black Elk
19th Century Sioux

When the moon is full, the seas rise up to reach it, sending wild waves of enthusiastic welcome. Oysters spread their shells wide, stretching to swallow it whole in the same way that they one day may slide down someone's slippery throat. Wolves howl at it, ears pricked, eyes glued adoringly on the object of their attention. Heads thrown back in ecstasy, they sit up very straight like any good dog and sing to it songs of atavistic refrain.

Sea creatures such as grunions and palolo worms which inhabit tube-like burrows on the sides of submerged coral reefs send their eggs up to the surface of the water at the precise hour of the full moon, whether the moon is visible or not. These and other moon-influenced creatures respond to the lunar rhythms in exactly the same way even when they are kept far from the tides in aquariums.

People, with their internal waters stirred like the tides, also turn to the moon. This is especially evident when the moon is full. The visage of that shimmering silver ball set in the vast blue sky, and the lunar-lit atmosphere it creates, has always held a universally mesmerizing, magnetic, mysterious, magical appeal.

MANY SOLEMN NIGHTS
BLOND MOON, WE STAND AND MARVEL...
SLEEPING OUR NOONS AWAY.

Matsunaga Teikoku
17th Century Japanese

The full moon is considered to be the climactic time of the moonth. The period of fulfillment, the blossoming of the expectation, the promise of the new beginning moon fulfilled. Mohammed exclaimed, "Verily ye shall see your Lord as ye see the moon on the night of its fullness without confusion in the vision of him." Young people in Zaire stay up throughout the "white nights" when the moon is full making music and dancing in her light.

The Hebrew word, *sabbath* is from the Babylonian, *sabattu*, meaning full moon. The Christian Easter is always on the first Sunday after the first full moon after the Spring Equinox, and the Jewish Passover, the full moon closest to it. Witches, practitioners of the old Wiccan religion, still convene on the full moon at which time they ceremonially draw it down to Earth for the purpose of using its energy to create positive magic.

THE MOON APPEARED IN ALL HER FULLNESS
AND SO THE WOMEN STOOD AROUND THE ALTAR

Sappho of Lesbos
7th Century B.C. Greek

In Greece, women used to present the moon goddess Artemis with a lighted full moon cake in honor of her monthly birthday. These moon offerings were the original birthday cakes. Women in the Orient still bake moon cakes in honor of their moon goddess, Ch'ang O. These round sweet rice balls are enjoyed at the time of her birthday on the full Harvest Moon in the autumn. The popular festivities, celebrated throughout Asia to this day, are usually in the form of moon watching parties.

CLOUDS COME FROM TIME TO TIME
AND BRING MEN A CHANCE TO REST
FROM LOOKING AT THE MOON.

Matsuo Bashö
17th Century Japanese

Phases of the Moon

Waning Moon

The moon is first seen as a sliver. Then, for a tad more than fourteen days it slowly expands. From right to left it inflates, fills out and becomes full of itself. Once it reaches its roundest form, the lunar disk reverses its monthly shape-shift, once again from right to left. This time, gradually shrinking in girth until it disappears altogether after fourteen-plus more days.

There are many explanations about what causes this fascinating waxing and waning phenomenon. The Aborigines of the Encounter Bay area of Australia tell of the excessive life style of the moon which causes her to gain and then lose weight. The Klamath people of Oregon say that the moon is falling to pieces as it wanes.

According to the Dakotas, the full moon is nibbled away by hungry mice. Once they have consumed it entirely, a new moon is born and will grow only to succumb in the same fashion. In the Balkans, it is wolves who devour the round moon. This is the story of record throughout the Americas, featuring different marauding animals in each bio-region.

In ancient Egypt, the waning moon was believed to be the only visible remains of the god Osiris who had been destroyed by his enemies. Early Semitic people held that the moon was under siege by seven demons who gradually broke it down. The Wongibon folk of New South Wales saw the old crescent moon as the bent back of a crippled old man.

After the hope and expectancy experienced at the new and first crescent moon, the growth and development of the waxing stage and the climactic charge of the full moon, the waning moon can seem sad, eliciting feelings of longing and loss — a reminder of the passage of time, of youth. Perhaps this explains why there is such a dearth of waning moon celebrations worldwide.

WANING AND WASTING AWAY
THE MOON DISAPPEARS —
HOW COLD A NIGHT!

Anonymous
Japanese

But waning is not a negative, downward descent. There is a critical, valuable place in the cycle for the waning process. Waning is less a matter of fading, than of distillation. Of ingathering. Of seasoning. The shriveled crescent of the old moon is like the intensified sweetness of a once succulent fruit that has been slowly dried in the sun.

The new moon is the arbor, the full moon is the grape and the waning moon is the wine (stored in the dark moon cellar). The waning of the moon suggests a certain maturity. Like the harvest, it is the collection of experience, of perspective, of the wisdom to know what to preserve for the future based on the cycles of past experience.

O LADY MOON YOUR HORNS POINT TOWARD THE EAST;
SHINE, BE INCREASED:
O LADY MOON, YOUR HORNS POINT TOWARD THE WEST;
WANE, BE AT REST.

From Sing-Song
Christina Rossetti
19th Century English

The waning moon is associated with the Grandmother Goddesses, the crones — at once all-wise and benevolent, demanding and devouring — like Macha Alla of the steppes of Central Asia, the Sumerian Inanna, Hecate of Greece, the Hopi Hard Beings Woman and the Maori Moon Mother who eats the dead. People everywhere who hunt, gather, grow, herd, fish or forage for their food understand that something has to die for something else to live. Death feeds life. And so the cycles continue.

That you are lofty as heaven—
    be it known!
That you are broad as the earth—
    be it known!
That you devastate the rebellious land -
    be it known!
That you roar at the land—
    be it known!
That you smite the heads—
    be it known!
That you devour cadavers like a dog—
    be it known!
That your glance is terrible—
    be it known!
That you lift your terrible glance—
    be it known!
That your glance is flashing—
    be it known!

(That, oh my lady, has made you great,
    you alone are exalted!)

Oh my lady beloved of An,
    I have verily recounted your fury!

From The Exaltation of Innana
Princess Enheduanna
3rd Century B.C. Akkadian

Moon Phases

Dark/New Moon

))))●●●●●((((

You Moon! Have you done
something wrong in heaven,
That God has hidden your face?

Jean Ingelow
19th Century English

If the presence of the moon stimulates our senses and activates our spirits, its absence creates a feeling of foreboding, of sadness. The Sumerians called the dark days of the moon "the days of lying down." East Indians call this time the "inside days." In Polynesia, people know that our silver satellite has gone back home to take a siesta. The Aborigines say that the Moon Woman has to recuperate from too much partying. Some cultures say the moon is dying during its decrease and disappearance. Others designate moonless nights as "the naked time."

Cultures everywhere regard the phase when the moon is out of sight as a precarious period during which we must exercise extreme care in all undertakings. The dark moon was associated with nefarious acts among the Canaanites, Hindus, Jews and Moroccans. The Babylonians neutralized the negative effects of these "black nights" by fasting and following a ritual program of propitiation.

In Europe, protective, pacifying, lucky moon charms were made from the feet of rabbits killed during the dark moon. The very best results could only be obtained by using the left hind foot of a rabbit slaughtered in the dark of the moon in a graveyard by a person with crossed eyes.

HA GO WAY NAH U NA
HA GO WAY NAH U NA ...
WE WAIT IN THE DARKNESS!
COME, ALL YE WHO LISTEN,
HELP IN OUR NIGHT JOURNEY:
NOW NO SUN IS SHINING;
NOW NO STAR IS GLOWING;

COME SHOW US THE PATHWAY:
THE NIGHT IS NOT FRIENDLY;
SHE CLOSES HER EYELIDS;
THE MOON HAS FORGOT US,
WE WAIT IN THE DARKNESS.

Darkness Song
Seneca Nation of the Iroquois

If the monthly dark period of the moon affects us so deeply, so strongly, imagine the influence of an eclipse of the moon (which is only possible when the moon is full). What if the moon didn't come back? And where does it go, anyway?

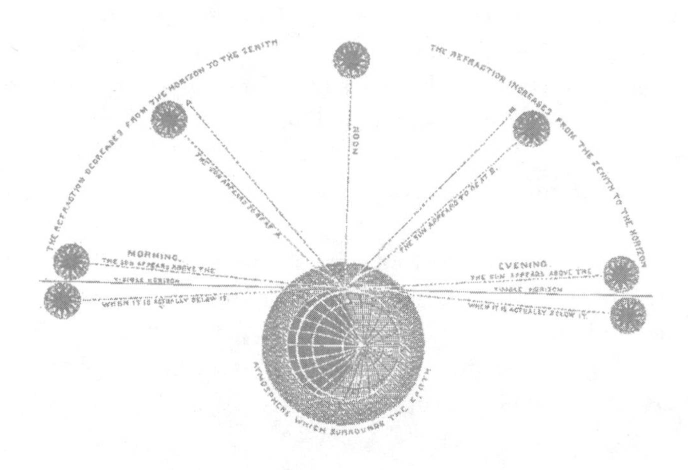

COME OUT COME OUT COME OUT
THE MOON HAS BEEN KILLED

WHO KILLS THE MOON? CROW
WHO OFTEN KILLS THE MOON? EAGLE
WHO USUALLY KILLS THE MOON? CHICKEN HAWK
WHO ALSO KILLS THE MOON? OWL
IN THEIR NUMBERS THEY ASSEMBLE FOR MOON KILLING

COME OUT, THROW STICKS AT YOUR HOUSES
COME OUT, TURN YOUR BUCKETS OVER
SPILL OUT ALL THE WATER DON'T LET IT TURN BLOODY YELLOW
FROM THE WOUNDING AND DEATH OF THE MOON.

O WHAT WILL BECOME OF THE WORLD, THE MOON
NEVER DIES WITHOUT A CAUSE
ONLY WHEN A RICH MAN IS ABOUT TO BE KILLED
IS THE MOON MURDERED.

LOOK ALL AROUND THE WORLD, DANCE,
THROW YOUR STICKS, HELP OUT, LOOK AT THE MOON,
DARK AS IT IS NOW, EVEN IF IT DISAPPEARS
IT WILL COME BACK, THINK OF NOTHING
I'M GOING BACK INTO THE HOUSE...AND THE OTHERS WENT BACK.

Moon Eclipse Exorcism
Anonymous Alsean

The moon never retreats for long. She is always, inevitably, drawn back to Earth by its inexorable pull. And so is the moon, eternal in her rebirths and resurrections, a certainty that humanity has grown to trust. Within the darkness of her disappearance and death, germinates the seed of her return. The dark moon is thus the alpha as well as the omega, the beginning anew which is born of the death of the old.

Moon Struck

By the Light of the Silvery Moon

Unlike the sun that dawns each day, the moon comes and goes, making it seem less consistent, more erratic and elusive. The longest period during which we who live in a temperate climate are separated from the sun is the length of the winter night. But, as interminable as that might sometimes seem, the breaking dawn always brings back the sun. Each morning, no matter how bleak, solar contact is renewed, and along with it, our sense of the sun as our constant companion.

Oddly, the ancients didn't necessarily associate the sun as the source of light. Genesis tells us that daylight was created before the sun and the moon. Many tribal peoples venerate the moon for its light-giving capacity. They consider moonlight to be more important because it illuminates the night when it is more needed, whereas, the sun only shines in the daytime when the sky is light already! The Desana people of Columbia credit the "night sun" with promoting fertility through the production of light/life-giving dew.

> No sun art thou that blinds and that inspires,
> And bleeding ends after a life in fire–
> Thou art what to the ailing singer his poem,
> Distant, but oh! the mild, mild light of home.
>
> *Annette Elizabeth von Droste-Hulshoff*
> *18th Century German*

The moon's light, reflecting the sun's as it does, is far less intense. Plutarch remarked that, "The effects of the moon are similar to the effects of reason and wisdom, whereas those of the sun appear to be brought about by physical force and violence." We can stare at it, and do — some of us for hours at a time — without suffering the damaging effects of the sun. Of course, it is the indirect light of the sun that we see anyway when we look at the moon — a mirror image indelibly imprinted in the eye of the beholder. It is interesting to note that moonlight dramatically affects our perception of the world. Our peripheral vision is vastly improved in moonlight, offering us a more expansive nocturnal view.

QUEEN OF THE SILVER BOW! BY THY PALE BEAM,
ALONE AND PENSIVE, I DELIGHT TO STRAY,
AND WATCH THY SHADOW TREMBLING IN THE STREAM,
OR MARK THE FLOATING CLOUDS THAT CROSS THY WAY;
AND WHILE I GAZE, THY MILD AND PLACID LIGHT
SHEDS A SOFT CALM UPON TROUBLED BREAST:
AND OFT I THINK FAIR PLANET OF THE NIGHT—
THAT IN THY ORB THE WRETCHED MAY HAVE REST;
THE SUFFERERS OF THE EARTH PERHAPS MAY GO—
RELEASED BY DEATH - TO THY BENIGNANT SPHERE;
AND THE SAD CHILDREN OF DESPAIR AND WOE
FORGET IN THEE THEIR CUP OF SORROW HERE.
OH THAT I SOON MAY REACH THY WORLD SERENE,
POOR WEARIED PILGRIM IN THIS TOILING SCENE!

Charlotte Smith
18th Century English

Silver has long been regarded as the color of the moon. Silver was one of the earliest metals known to humans and used for money from the beginning. In its pure state it is quite white with a strong metallic luster, very moon-like. The Hindu moon deity, Chandra, is called the White or Silvery One. This age-old association survives in the terminology of modern science. The chemical term for fused nitrate of silver is lunar caustic.

Cool, clear lunar light washes the world clean with a crystalline vision, revealing what has been hidden. Silver is thus symbolic of spiritual illumination. For this reason, silver, the metal of the moon is often used for ritual paraphernalia and tools. Silver was used exclusively in Incan moon worship, as gold was used in supplication to the sun. The altars and shrines in the Temple of Diana at Esphesis were created from silver.

THE MOON IS SILVER WITHOUT AND A JEWEL WITHIN,
COOL IN BOTH ASPECTS, INNER AND OUTER.

Buddhist Saying

Genesis 44:5 tells of Joseph's vision cup which was made of silver. As in all mythology, the chalice represents the lunar source of the "waters of enlightenment." The Sufi philosopher, Ibn al-Fârid said, "Our cup is the full moon, the wine the sun." By Mohammed's order, all religious amulets must be made with silver, the moon's special metal.

During the Middle Ages, silver coins and totems were used in lunar worshipping ceremonies by otherwise Christian women who retained their trust in the benevolence of the Moon Goddess. During the inquisition, silver charms were used by Christians to repel witches. We still divine our future with silver (alloy) when we toss a coin to make a heads or tails decision.

GLEAMING IN SILVER ARE THE HILLS!
BLAZING IN SILVER IS THE SEA!

AND A SILVERY RADIANCE SPILLS
WHERE THE MOON DRIVES ROYALLY!

CLAD IN SILVER TISSUE, I
MARCH MAGNIFICENTLY BY!

"Washed in Silver"
James Stephens
19th century English

Moon Struck

The Moon and Me

THE MOON IS A DIFFERENT THING TO EACH OF US.

Frank Borman
Apollo VIII Astronaut
December 24, 1968

Long ago, humankind adopted the moon as our kin. As the moon reflects the sun, it seems to mirror our minuscule existence as well. When we gaze upon its countenance, we see our own face transmitted back through space, and, like Narcissus, we are enamored. We identify ourselves with its moody mutability.

THE GROWING AND DYING OF THE MOON REMIND US
OF OUR IGNORANCE WHICH COMES AND GOES

Black Elk
19th Century Sioux

We recognize its influence. We mimic its magical monthly modification and apply its lessons of transformation to our own devices. Celestial shape-shifter, its cycles seize our imagination and inspire us to change by its lofty example. By following the moon's progress, we can chart our own.

People have always engaged the moon in a most intimate and personalized relationship. What child hasn't marveled that the moon follows her when she rides or walks at night?

I SEE THE MOON, AND THE MOON SEES ME.
GOD BLESS THE MOON AND GOD BLESS ME,

Traditional Nursery Rhyme.

I was there. Staring at the blue crystallized moon.
The icy night wind blew me across the city. Before I knew
it I was walking on the sea. I felt as light as a feather,
but I was still there. Staring at the blue crystallized moon.
The icy night wind blew me across the sea. Before I knew it
I was floating in heaven. I felt as light as a feather but I
was still there. Staring at the blue crystallized moon. The
icy night wind blew me out of heaven. Before I knew it
I was standing there before the devil. I felt as light as
a feather. But I was still there. Staring at the blue
crystallized moon. The icy night wind blew me out of
Hell. I flew out of the world. Then I woke up. But I was
still there. Staring at the blue crystallized moon.
As it was staring at me.

Jonathan Chang, 5th grade
20th Century American

We identify with the moon on a deep emotional level and want to make it all our own. It is almost as though we can each hear it speak to us in a private lunar language which only we can understand. When the moon calls to us, we answer from the very center of our emotional and spiritual being.

I CANNOT SLEEP
FOR THE BLAZE OF THE FULL MOON.
I THOUGHT I HEARD HERE AND THERE
A VOICE CALLING.
HOPELESSLY I ANSWER "YES."
TO THE EMPTY AIR.

Lady Tzu Yeh
3rd Century Chinese

Although the moon is almost a quarter of a million miles away, its influence is immense. Its gravitational force is 46% stronger than that of the sun. And at its two extreme positions, when it is full and when it is dark, we experience an intensely potent psychological and physical reaction, as if it were exerting the same pull on our emotions as it does on the waters which surround us.

THE WAVERING PLANET MOST UNSTABLE,
GODDESS OF THE WATERS FLOWING,
THAT BEARS A SWAY IN EACH THING GROWING
AND MAKES MY LADY VARIABLE.
OFT I SEEK TO UNDERMINE HER,
YET I KNOW NOT WHERE TO FIND HER.

"The Wavering Planet" from
Giles Farnaby's Canzonets to Fowre Voices
Anonymous, 16th Century English

As Pliny, the Greek natural historian noted, "The moon draws the sea after her with a powerful suction." So too, are we sucked into that impressionable wash. We are quite conscious of its circuits, and we tend to relate them to our own earthly psyches and to our internal cycles of growth and decline, expansion and contraction.

CORRESPONDENCE:
WHEN I HAVE SAD THOUGHTS
EVEN THE MOON'S FACE
EMBROIDERED ON MY SLEEVE
IS WET WITH TEARS.

Lady Ise
10th Century Japanese

From fullness, the moon decreases in size before our very eyes, eliciting a certain sadness as it shrinks, departing daily in slender degrees, like slices cut from a voluptuous fruit. It changes its guise steadily, until it finally disappears altogether for three dark, depressing days at a time, after which it returns again in reversed increments, producing an exalted sensation of growing expectation and excitement along with its expanding girth.

We make the moon our personal friend, our confessor, our reference, our partner. We tell it our secrets and invite its soft light to illuminate our inner feelings. We find comfort in its predictable and orderly progress through the cycles, reminding us that everything is mutable, that everything changes and that this mood, too, shall pass.

From darkness
I go onto the road
Of darkness.
Moon, shine on me from far
Over the mountain edge.

Lady Izumi Shikibu
10th Century Japanese

The light can visionary thoughts impart,
And lead the Muse to soothe
the suffering heart.

Helen Maria Williams
19th Century French/English

To this day, it is a customary practice in Japan to go outside in the moonlight when one is troubled. The procedure is to bow three times, meditate until calm, and then bow three more times with reverence to the Lady in the Moon when your spirit has been healed.

Moon Struck

MOON MADNESS

The concept of moon madness is an old and widespread one, confirmed in language. Moon and mind and spiritual power are linked etymologically from the Indo-European precursor of both the Sanskrit, *manas* and the Latin, *mens*. From this root are derived the English words: "menstruation," moon blood; "mania," moon madness; and "numinous," moon magic.

Two of the names of the Roman moon goddess were Luna and Mana. Her devotees, lunatics and maniacs, were condemned by the Church as mad, crazy, or "silly" — a word which had formerly meant blessed.

WYFE, I WEENE THOU ARET DRONKE OR LEUNITIKE.
NAY HUSBAND, WOMEN ARE NEVER MOON SICKE.

19th Century English Epigram

The traditional forms of lunar worship were banned by the Church and all things associated with the moon were held to be evil and wrong, having a detrimental affect on people, especially on women who are connected to the moon by their own cycles. Saint Augustine denounced women for dancing "impudently and filthily all the day long upon the days of the new moon," even as their Hebrew sisters were scorned for wearing lunar amulets by the biblical prophet in Isaiah 3:18.

CHERISH YOUR MADNESS
MY SISTER, BUT
DISCRETELY O DISCRETELY

WALK WARILY IN THE
WORLD KEEP ONE
EYE ON THE MOON
FOR SAFETY

AND IF DISCRETION FAILS
MY SISTER, PUT
YOUR MADNESS IN
A BAG AND

RUN LIKE HELL.

"Tactical Advice"
Barbara Starrett
20th Century American

Paracelsus, famed physician of the Middle Ages, considered the brain to be a "microcosmic moon" which responds to the full moon with an unsettled, turbulent state of mind. He said that the moon had the "power to tear reason out of a man's head by depriving him of humors and cerebral virtues."

Shakespeare had Othello say, "It is the very error of the moon, she comes more nearer Earth than she was wont, and makes men mad." For this reason, it was considered dangerous to sleep in the light of the moon — especially the full moon. Ironically, a sort of lunar therapy is applied in Central Asia where it is still believed that the moon's reflection on water is the prime remedy for nervous hysteria.

In 1842, the British Parliament passed the Lunacy Act which defined a lunatic as a deranged individual who was rational and lucid, that is, sane, during the dark and waxing phases of the moon, but "afflicted with a period of fatuity in the period following after the full moon." As recently as 1940, an English soldier accused of murder used "moon madness" as his defense. He claimed he was overtaken each month at the full moon.

BETTER TO SEE YOUR CHEEK GROWN HOLLOW.
BETTER TO SEE YOUR TEMPLE WORN,
THAN TO FORGET TO FOLLOW, FOLLOW,
AFTER THE SOUND OF A SILVER HORN.

BETTER TO BIND YOUR BROW WITH WILLOW
AND FOLLOW, FOLLOW UNTIL YOU DIE,
THEN TO SLEEP WITH YOUR HEAD ON A GOLDEN PILLOW,
NOR LIFT IT UP WHEN THE HUNT GOES BY.

BETTER TO SEE YOUR CHEEKS GROWN SALLOW,
AND YOUR HAIR GROWN GRAY, SO SOON, SO SOON,
THAN TO FORGET TO HALLO, HALLO,
AFTER THE MILK-WHITE HOUNDS OF THE MOON.

"Madman's Song"
Elinor Wylie
20th Century American

Any fireman, cop, teacher, bartender or emergency room worker will be able to regale you with tales of the eccentric and/or violent behavior which clearly occurs in coincidence with the new and full moon. According to Harvey Schlossberg, director of the Psychological Services Unit of the New York Police Department, "There's really no way to explain it scientifically, but there is an increase in the number of assaults and crimes between people at the full moon." The best known research to document the relationship between the murder rate and the phases of the moon was conducted by Dr. Arnold L. Lieber, a psychiatrist, and his associate, Carolyn Sherin, a clinical psychologist. A study of 4000 homicides which occurred between 1956 and 1970 in Miami, and between 1958-1970 in greater Cleveland showed a "statistically significant correlation" between murder and the cycles of the moon over a fifteen and a thirteen year period. (1)

Homicides were shown to peak twice monthly, at the full moon and just after the new moon. Son of Sam, New York City's most infamous mass killer, committed eight murders between July 29, 1976 and July 31, 1978. On five of these nights the moon was either full or new.

Julius Caesar, Jesus Christ, Alexander II of Russia, Leon Trotsky, King Abdul ibn Hussein of Jordan, Mexican President Francisco Madero and Dominican dictator Rafael Trujillo Molina were all assassinated or executed when the moon was full. The My Lai massacre was committed in the full silver stage of the moon. And the worst sports disaster in history happened when 328 people were trampled to death in a riot at a soccer match under the light of the full moon.

Suicides seem to follow the same pattern. By studying the records kept by the Cuyahoga County coroner's office, Paul and Susan Jones were able to identify clusters of suicides occurring at the full and new moons. (2) More than one hundred young people tried to kill themselves on the same full moon day in Tehran, Iran in 1978. Many succeeded. In the same year, it was reported that nine out of the 23 suicides committed that year from the Golden Gate Bridge were on the full moon.

> MY EYES WERE LIKE THE LOTUS
> MY ARMS HAD THE GRACE OF THE BAMBOO
> MY FOREHEAD WAS MISTAKEN FOR THE MOON.
> BUT NOW...
>
> *Maturai Eruttalan Centamputan*
> *3rd Century Tamil*

Probably the most famous sort of lunar madness is lycanthropy, the conviction that the power of the full moon can cause a person to become a werewolf — doomed to grow fur and claws and wander at night, hunting, killing, cannibalizing, sleeping in cemeteries and howling at the moon. This notion has been quite widespread throughout history. Apparently, the biblical Babylonian King Nebuchadnezzar suffered from such extreme depression that he felt himself turn into a werewolf.

> EVEN A MAN WHO IS PURE IN HEART
> AND SAYS HIS PRAYERS BY NIGHT
> CAN BECOME A WOLF WHEN THE WOLFBANE BLOOMS
> AND THE MOON SHINES BRIGHT.
>
> *Traditional English Folk Song*

A rare modern such case was written up in the Canadian Psychiatric Association Journal.(3) The patient let his beard and hair grow and genuinely believed that it was fur and that he was a wolf in the light of the moon. The doctors witnessed and tracked his attacks which always fell on the full moon. They could attribute no other organic explanation for his spells. Incidents of werewolfism occur even today in India, China, Africa and the traditional indigenous cultures of the Americas.

> OH! YE IMMORTAL GODS!
> WHAT IS THEOGONY?
> OH! THOU, TOO, MORTAL MAN!
> WHAT IS PHILOSOPHY?
> OH! WORLD, WHICH WAS AND IS, WHAT IS COSMOGONY?
> SOME PEOPLE HAVE ACCUSED ME OF MISANTHROPY;
> AND YET I KNOW NO MORE THAN THE MAHOGANY
> THAT FORMS THIS DESK, OF WHAT THEY MEAN–LYCANTHROPY
> I COMPREHEND, FOR WITHOUT TRANSFORMATION
> MEN BECOME WOLVES ON ANY SLIGHT OCCASION.
>
> *Lord Byron*
> *19th Century English*

Moon Struck

Moon as Muse

Ah, Moon of my Delight who know'st no wane,
The Moon of Heav'n is rising once again.

Omar Khayyâm
11th Century Persian

The moon and the earth are like lovers, locked by gravity in an unending, inseparable embrace. Their unbreakable bond — despite, or perhaps because of, their occasional separations and inevitable reunions — is a recognizable model for earthly love. Stimulated by their teasing, loyal dance, the erotic pairing of these celestial bodies has excited couples through the ages.

Look you! How strangely wan and pale
The moon climbs up to the mountain height!
When over the summit it sets its sail
It brings the night
It brings the night—and our love's delight.

Lady Yukada
8th Century Japanese

The shimmering moon lights the way for romantic associations, passionate pairings, secret assignations. Moonlight ignites the passions and stimulates the longings, the poignant yearnings of the emotions. What lonely lover, separated from the subject of her affections, does not take heart at the romantic revelation that the moon she sees is the same moon gazed upon, perhaps even simultaneously, by her distant love?

Someone else
looked at the sky
with the same rapture
when the moon
crossed the dawn.

You told me it was
because of me
you gazed at the moon.
I've come to see
if this is true.

Lady Izumi Shikibu
11th Century Japanese

The moon is described in many cultures as a highly charged sexual being. When Chinese calendars were entirely lunar, they were divided into "houses" each of which was occupied by one of the moon goddess's warrior-hero consorts. It was understood that Lady Moon spent each night of the moonth in a different celestial mansion with a different lover.

The moon rises
Stealing the Sun's light
Between her thighs
The man steals the nectar
between her thighs.

Folk Song
Chhattisgarh India

There are many widespread stories about the moon falling in love with a human. In Papua New Guinea it is said that a young girl marries the moon who de-flowers her. Her monthly blood is proof of their coupling. In the Cook Islands of Oceana it is told that the Moon fell in love with a pretty young woman on Earth. He swept down from the sky to visit her and eloped with her back to the sky. You can still see her up there with a pile of leaves for her oven and her tongs to poke the embers.

ENDYMION THE SHEPHERD,
AS HIS FLOCKS HE GUARDED, SHE, THE MOON, SELENE
SAW HIM, LOVED HIM, SOUGHT HIM.
COMING DOWN FROM HEAVEN
TO THE GLADE AT LATMUS,
KISSED HIM, LAY BESIDE HIM.
BLESSED IS HIS FORTUNE.
EVERMORE HE SLUMBERS,
TOSSING NOT NOR TURNING, ENDYMION THE SHEPHERD.

Theocritus
3rd Century B.C. Greek

The Ojibwa people who live in the Lake Superior region of the United States and Canada tell the following tale:

> There once lived a girl called Lone Bird, daughter of She Eagle and Dawn of the Day. Many men wanted to marry her, but Lone Bird wanted nothing to do with any of them to the great regret of her parents.
>
> One day while gathering maple syrup in her birch bark bucket, she sat down on a rock at the lake shore to think about her single life and contemplate her future. She realized that none of the animals lives alone. This made her very sad. Dejected, she sat for quite a long time.
>
> When she finally rose from her reveries, the full moon had laid a sheet of silver across the vast expanse of shimmering waters. "Oh how beautiful you are," she cried to the

moon. "If I had you to love, certainly I would never be lonely again." Moved by her deep longing, the Great Spirit carried her into the sky to join the moon.

When she didn't return, her frantic father searched everywhere for her. Finally he looked up, and there, cradled in the arms of the moon, was Lone Bird, smiling down on him contentedly.

The exquisite loveliness of the moonlit night has an entrancing effect. Its specter is somehow deeply inspiring to the senses, the creative spirit, the soul. Many moon goddesses such as Astarte, Isis, Ishtar, Selene, Venus, were also Goddesses of Love and Art, rulers of beauty and heart. The moon serves as muse to lovers, artists and poets, alike.

THE MOON LIKE A FLOWER,
IN HEAVEN'S HIGH BOWER
WITH SILENT DELIGHT
SITS AND SMILES ON THE NIGHT.

From "Night"
William Blake
18th Century English

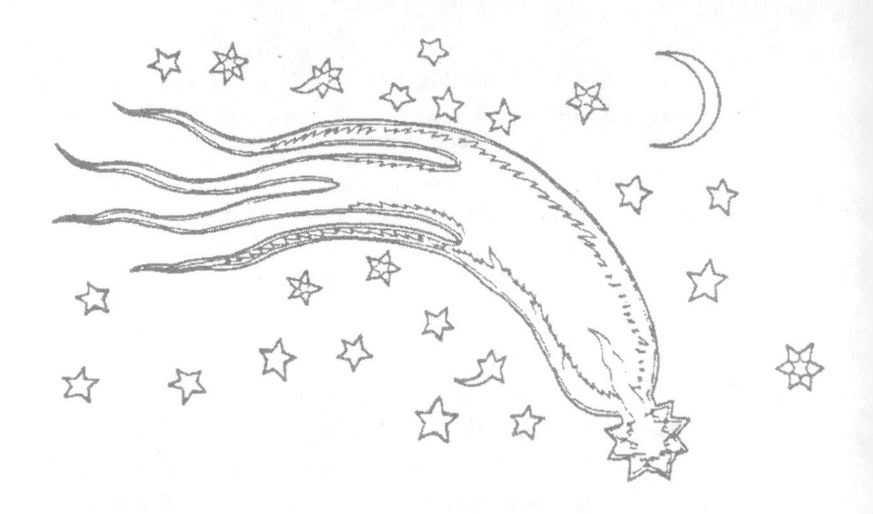

The soul, like the moon,
is new, and always new again.

And I have seen the ocean
continuously creating.

Since I scoured my mind
and my body, I too, Lalla,
am new, each moment new.

My teacher told me one thing,
Live in the soul.

When that was so, I began to go naked and dance.

"The Soul"
Lal Ded
14th Century Kashmiri

AND I WOULD BE THE MOON
SPOKEN OVER YOUR BECKONING FLESH
BREAKING AGAINST RESERVATIONS
BEACHING THOUGHT
MY HANDS AT YOUR HIGH TIDE
OVER AND UNDER INSIDE YOU
AND THE PASSING OF HUNGERS
ATTENDED, FORGOTTEN.

DARKLY RISEN
THE MOON SPEAKS
MY EYES
JUDGING YOUR ROUNDNESS
DELIGHTFUL.

From "On A Night Of The Full Moon"
Audre Lorde, 20th Century American

MOON SHINES ON VALLEY
GRASS SLEEPS BY RIVER
NOW WHY DON'T YOU COME
SIT DOWN WITH ME
& LOVE ME A LITTLE
AS I LOVE YOU.

Romany Gypsy Song

Watching alone by the ancient city wall,
Thinking of one who was too beautiful,
What did I see? What did I hear?

Moonlight, quivering over empty courtyards,
A voice calling out of the midnight shadows.
One name, her name, echoes across the silence.
Light feet, her feet, in shoes of peacock feathers,
Dance through the empty halls.
Will they never rest?

Thinking of joys that ended and sorrows
which never end
I find my white robe spangled with tears for her.

Anonymous
12th Century Korean

The Owl and the Pussycat went to sea
In a beautiful pea-green boat;
They took some honey, and plenty of money
Wrapped up in a five-pound note.
The Owl looked up to the stars above,
And sang to a small guitar,
"O lovely Pussy! O Pussy, my love,
What a beautiful Pussy you are,
You are,
You are!
What a beautiful Pussy you are!"

. . . They dined on mince, and slices of quince,
Which they ate with a runcible spoon;
And hand in hand, on the edge of the sand,
They danced by the light of the moon,
The moon,
The moon,
They danced by the light of the moon.

"The Owl and the Pussycat"
Edward Lear, 19th Century English

Living in Sync with the Cycles

The eight phases of the moon can be symbolized by the life cycle of plants. The dark new moon is the seed, the germination. The first crescent is the sap, the sprout; the waxing, the bud. The waxing gibbous moon is the foliage; the full, the fully opened flower. The waning gibbous is the fruit, the harvest; the waning, the chaff, the dross. The last crescent is the dried fruit, the jam, the relish.

What applies to plants, also applies to people — the fertilization, germination, birth, growth, reproduction, maturation, physical decline and death stages of human beings seem to be reflections of those of the moon. Because of this confluence, people have long conducted the celebrations of the seasons of their own lives with the cyclical movement of the moon. The moon rules our most personal moments of supreme import: birth, fertility, death.

The moon, widely regarded as the domestic deity, also rules the mundane — the real-life chores of diligence, discipline and maintenance. Over the ages, folk wisdom has developed an extensive repository of moon lore to guide the daily activities of personal health and hygiene, domestic chores and secular activities.

New Moon

The new moon is a time of initiation and new beginnings. It is the time to plant the seed of what you want to grow in your life. Making a wish, an intention, a promise, a vow on the new and new crescent moon is a common custom. As the moon grows, so will the wish. It will come true by the time the moon completes its cycle and returns full circle to new again.

The new moon lends energy to the intention to rout out all the negative habits in one's life. It aids one to stop self-destructive behaviors like smoking, drinking too much alcohol or coffee, or eating too much. It helps one to start afresh on the right foot and initiate positive change. Consequently, it is a propitious time to begin a new project or discipline; to start a diet or an exercise regime.

This is the right time to begin a courtship, a new business, a trip, a diary or any new venture that you want to develop and to prosper. If you count your money now, it will multiply as the moon grows. In New Guinea, women pray to the new moon to protect their husbands while they are traveling away from home on hunting expeditions, and to prevent them from disappearing like the moon.

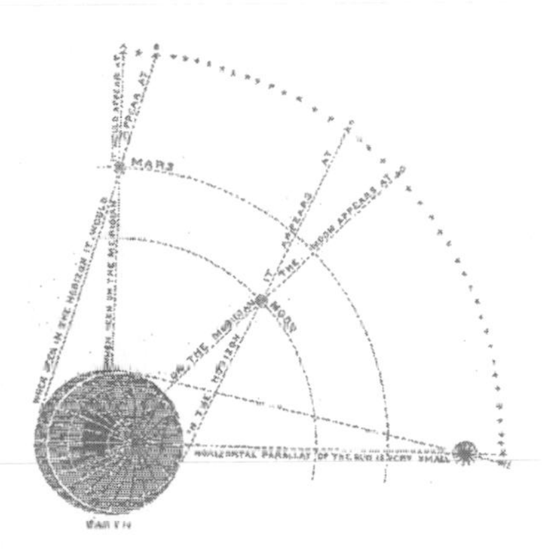

Waxing Moon

The stage from the new moon to the full moon is one of increasing development, growth and expansion. Many folk, including Estonians, Finns, Yakuts and Orkney Islanders marry only in the hopeful phase of the waxing moon, symbolizing the wish for a happy and bountiful union to come.

The moon has long been seen as the midwife protector of mothers. Plutarch has written that the moon has a "special hand in the birth of children." Labor and child birth are frequently thought to be easier in the fourteen day period of the waxing moon.

In Scotland, babies are weaned from the breast during the waxing moon so that the child can grow with the moon and flourish. In Lithuania, baby boys are weaned at the same time for the same reason, but girl children are weaned at the waning moon to discourage growth.

Full Moon

The full moon represents the fullest potential realized; fertility fulfilled. The fat round moon symbolizes a full belly, a full pantry, a full purse, a full womb, a full life. The Gaelic word for full moon, *Gealach*, is the root for the word that means good fortune. It is widely considered to be especially lucky for romance. Perhaps this is because a greater percentage of women ovulate with the full moon than at other times.

The Ancient Greeks, Celts and German Jews during the Middle Ages traditionally married only when the moon was full. In Euripides, when Clytemnestra asks Agamemnon when he would give the hand of Iphigenia to Achilles in marriage, he replies. "When the full moon comes forth with good luck."

It has long been thought that children born in the full light of the moon will be more hearty. Francis Bacon wrote in his 18th Century *Natural History*, "It may be that children and young cattle that are brought forth in the full of the moon are stronger and larger than those that are brought forth in the wane."

According to the Celts, babies born at this time bring fortune for the same reason. The Baganda people of central Africa bathe their first born child by the light of the first full moon after its birth to bless it with health and wealth.

In a study of over half a million births by Walter and Abraham Menaker, they discovered that the birth rate was significantly higher in the three day period surrounding the full moon than at any other time of the month.

On the domestic front, the old wives counsel not to wash clothes for the first time in the full moon, or else they will not last very long. On the other hand, leaving a stained table cloth outside in the moon-light will whiten it. The fullest and fluffiest featherbeds are stuffed at the full moon.

Waning Moon

Farmers and ranchers have long refrained from planting and slaughtering while the moon wanes from full to new, for fear that the crops, the stock, would, too, wither and die.

Since the energy of the moon is that of withdrawing, drying, closing, it is not considered to be an auspicious time to be born. It was once thought that children born from the wane until the dark of the moon would be prone to poor health. In Cornwall, the expression, "No moon, no man," reflects the belief that a child born now will not live to see puberty. In fact, current science has determined that the fewest births in the month occur at the dark moon.

This is not the time to start anything important. Superstition says that if you move your residence now, it bodes ill for your happiness there. If you wear a new garment, it will not prove to be durable.

Since the waning phase is about retreat, it is said to be the best time for cleaning and washing, especially laundering linens. Everything works better, faster, and with less effort. The dirt departs with the moon.

Working with the Energy of the Moon

The moon has long been seen to rule the realms of water, and as such, she is the mother and mistress of all growing things; animal, mineral and especially vegetation. The *Yasht*, an Iranian text, says that plants are caused to grow by the heat of the moon. Several tribes in Brazil credit the moon with mothering the grasses. Grass grew on the moon according to the ancient Chinese.

Agricultural people have long coordinated their farming and husbandry tasks with the phases of the moon in order to increase both the quantity and the quality of their yield. It is generally assumed that in the way of sympathetic magic, the growth and health of plants and animals respond to and imitate the moon. During the waxing moon, the moisture on Earth waxes with it making the plants and animals active and juicy. When the moon wanes, living things slow their pace and dry out. "As above, so below," says the Jewish *Talmud*.

The general rule is that when the moon is new and waxing, it lends energy to beginnings, growth and increase, especially in plants that produce food or flowers above ground. When the moon is full, it is reflected in the ripeness and fertile, fullness of life on Earth, perfect for fattening, mating and breeding. When the moon wanes, it is time to plant plants that produce below the ground, root crops and bulbs. This is also the period to weed, prune, harvest. When the moon is dark, it is time to dig in, to mulch, to germinate.

Sowe peason and beanes, in the wane of the moone,
Who soweth them sooner, he soweth too soon,
That they with the planet may rest and arise,
And flourish, with bearing most plentiful wise.

From "Five Hundred Points of Good Husbandry"
Thomas Tusser, 16th Century English

New (Dark) Moon

- Plant potatoes so that they'll stay down deep
- Pick apples to keep them from rotting
- Put eggs under a hen for a brood of strong chicks
- Catch fish which now respond easier to the lure
- Erect fence posts for greater durability
- Spray fruit trees
- Gather fresh dew

Waxing Moon

- Plant/transplant annuals which will bear fruit above the ground
- Plant cereals for faster germination
- Sow lawns and other large areas
- Put down sod
- Transplant and re-pot house plants
- Harvest foods for immediate consumption
- Shear sheep
- Slaughter game and other meats
- Dowse for water

Full Moon

- Plant seeds during droughts
- Harvest grapes for wine
- Pick mushrooms
- Plant berries
- Plant trees
- Pick herbs for medicinal use
- Add fertilizer
- Catch crabs, clams and shrimp
- Mate/inseminate animals

Waning Moon

- Plant/transplant biennials, bulb and root-crop plants
- Prune bushes and trees
- Mow lawns
- Make hay
- Kill weeds
- Spread mulch
- Harvest food for preserving and storing for future use
- Dry beans, can vegetables, make jams and jellies
- Salt meat and fish
- Strain sediment from olive oil
- Gather up grain from the threshing floor
- Cut timber — in France such wood is labeled "preferred"
- Build with wood
- Excavate, do earth moving, work on cellars, lay stone
- Cut peat for less smoky fires
- Castrate and dehorn animals for less bleeding

MOON GARDEN

Moon gardens are cultivated specifically to be viewed by moonlight. They are usually composed of arrangements of all-white flowers, and feature special night blooming, fragrant species.

A typical moon garden might include white varieties of:

Alyssum
Asters
Baby's Breath
Columbine
Cosmos
Daffodils
Delphiniums
Japanese Iris
Moonflowers
Mums
Nicotiana
Night-Scented Stock
Petunia
Sweet Woodruff

Moon Wise

Healing Plants of the Moon

According to the old wives' wisdom of various folk traditions, the following common plants are ruled by the moon and are, as such, considered to be especially efficacious in their curative properties:

Plant	Folk Remedies
Camphor	*Wards off colds; repels unwanted lovers;induces sleep.*
Cucumber	*Because of its many seeds, it enhances fertility; cures headaches.*
Eucalyptus	*General healing tonic; take the green pods for colds and sore throats.*
Gardenia	*Attracts lovers.*
Lettuce	*Juice rubbed on forehead relaxes and induces sleep.*
Poppy	*Promotes fertility, prosperity and prophetic dreams.*
Sandalwood	*Burn to purify the air and create a healing, protective atmosphere.*
Succulents	*Aid in attracting love and abundance.*
Wild Tansy	*Use tea as mouthwash to relieve sore gums and toothaches; take with honey for sore throats; relieves menstrual cramps.*
Willow	*Helps in general healing, granting and accepting wishes and recognizing the blessings of the moon.*

Moon Medicine

Most cultures recognize the moon's influence on all living things, including the physical and emotional well-being of people, and they plan their healing regimes accordingly. The ancient Hebrews, Greeks, Romans and many African tribes all believed that the moon had a deleterious effect on health. The New Testament tells of a boy who is afflicted by seizures whom Jesus cures by banishing the evil spirit that has possessed him. The Latin *Vulgate Bible* describes this seizure-illness as *Lunaticus*, or moonstruck.

Epilepsy has long been thought to be a lunar affliction, flaring up when the moon becomes full. One old-time theory is that it is caused by excessive moisture on the brain brought about by the moon, the ruler of the waters, the rains and the tides. Interestingly, today's medical science supports this theory. It is now suspected that epilepsy is a result of cerebral edema, a critical accumulation of pressure from fluids surrounding the brain. Do these fluids rise like the tides at the full moon?

The concept of lunar sympathy links folk medicine to the moon in the same way that it does hunting, gathering, horticulture and animal husbandry. Not only do plants and animals (including humans) grow better in the waxing period of the moon, they thrive. They are stronger, more energetic, more resilient, more resistant to disease. And conversely, as the moon wanes, the living organism becomes weaker and more vulnerable to accident and illness.

New (Dark) to Waxing Moon

- The body absorbs and integrates during the period of the waxing moon. There is a tendency to gain weight, retain water and take in toxins. Consequently, it is especially important to expose your body to positive influences such as good nutrition, exercise, body work therapies and meditation.

- It is a good time to cut hair and nails to promote quick and healthy growth. The Roman emperor Tiberius would only visit his barber in the phase of the waxing moon for fear of going bald.

- In Staffordshire, England, the folk cure for whooping cough entails taking the child out to lie in the growing moonlight and exposing her belly to Mother Moon. Rub her tummy with your right hand while staring at the moon. Recite: "What I see, may it increase; What I feel, may it decrease." Obscure, but supposedly effective.

- Before the advent of breast enhancement surgery, the women of Naples, Italy would pray to the ripening round moon for a bigger bosom. Standing alone and naked, arms raised in petition, they intoned nine times, "*Santa Luna, Santa Stella, fammi crescere questa mammella*" (Holy Moon, Holy Star, make this breast grow for me).

- Gather medicinal plants, as they are now at their most potent.

Full Moon

- During the full moon and for several days before and after, the body, too, is brimming full. There is a dramatic tendency to hemorrhage now. Blood banks around the country report that their biggest demands come on the full moon and the two days following.

- Avoid surgery as there is danger of excessive bleeding. In India, it is considered proper medical procedure among traditional plastic surgeons to postpone all surgeries until after a full moon to avoid unnecessary scarring. Dr. Edson Andrews of Tallahassee, Florida conducted a four-year study of the incidence of serious hemorrhaging during surgery at Tallahassee Memorial Hospital. He discovered that 82% of bleeding emergencies occurred during the full moon. He noted "an amazing association of the full moon and the increased incidence of bleeding." He said that the evidence was so convincing that it threatened to turn him into a witch doctor and operate only on the dark days of the moon! (4)

Waning to New (Dark) Moon

- The body now releases, relaxes, rests. At this time, toxins and tensions are eliminated more easily, excessive water is expressed. It is easier during these two weeks to lose weight as well as addictions and other self-destructive habits. This is also the best time to fast.

- Cut your hair and nails if you want to discourage them from growing too quickly.

- A good time for having teeth filled and pulled.

- Dig for medicinal roots now.

- If you treat your warts or cut your corns now, they will not come back. The following is a letter to the editor of the British *Apollo* from the early part of the 20th Century:

> *Pray tell your querist if he may*
> *Rely on what the vulgar say,*
> *That when the moon's in her increase,*
> *If corns be cut they'll grow apace;*
> *But if you always do take care*
> *After the full your corns to pare,*
> *They do insensibly decay*
> *And will in time wear quite away.*
> *If this be true, pray let me know,*
> *And give the reason why 'tis so.*

Celebrating the Cycles

Lunar Worship

Worship of the moon was pandemic and in most cultures preceded a more recent shift of allegiance to the sun. The status of the moon has always held precedence in Asia, and all religious occasions are still computed on a lunar calendar.

Adam, according to Moses Maimonides, worshipped the moon in a manner which was a religious holdover from Babylonian times. Mt. Sinai, where Moses received the Ten Commandments, means Mountain of the Moon. Jews still venerate the new moon and celebrate its occurrence as a holy day, *Rosh Chodesh*. The crescent moon, symbol of the pre-Islamic moon goddess, Manat or Al-Lat, along with the morning and evening star Venus even now serve to represent Islam.

Women all over Europe, despite the dictates of the Church, continued to worship the moon up through the Renaissance, when women knew that if they wanted special succor they should pray not to God, but to the Lady Moon. This was especially true in Ireland and Iceland where women sought the eternal wisdom of the moon in sacred wells and, in some places, still do.

Even today, the most dapper, jaded urban dwellers will take an instant, astonished notice of the full moon as if seeing it for the very first time. In so doing, they celebrate a simple celestial observance which connects them with the cosmos if even for a single mystical moment. This awe-filled gaze is ceremony at its most elemental. Observance = observance.

The moon continues to reign supreme around the globe, her sovereign image ever widely, proudly proclaimed. The crescent moon, alone, or with one or more stars, is today displayed on the flags of Turkey, Pakistan, Tunisia, Libya, Algeria, Malaysia, Singapore, North Cyprus, Nepal, The Maldives and South Carolina. The moon displayed on the flag of Palau in the South Pacific is full and golden, glorious against a field of sky sea blue.

Celebrating the Cycles

Lunar Invocations

Welcome, precious stone of the night,
Delight of the skies. precious stone of the night,
Mother of stars, precious stone of the night,
Child reared by the sun, precious stone of the night,
Excellency of stars, precious stone of the night.

Gaelic

Moon, O Mother Moon, O Mother Moon,
Mother of living things,
Hear our voice, O Mother Moon!
O Mother Moon, O Mother Moon,
Keep away the spirits of the dead,
Hear our voice, O Mother Moon,
O Mother Moon! O Mother Moon!

Gabon Pygmy

O Light One.
O Bright One.
O Wheel of Ice
O Mirror Bright
We bow tonight.
Bless thou our rice!

Traditional Chinese

I shall prosper.

I shall yet remain alive.

Even if people do say of me,

"Would that he dies!"

Just like thee shall I do.

Again shall I arise.

Even if all sorts of evil beings devour thee.

When frogs eat thee up.

Many evil beings - lizards.

Even when those eat thee up.

Still dost thou rise again.

Just like you will I do in time to come.

Bo!

Takelma Tribe, Oregon

My brother the star, my mother the earth,

My father the sun, my sister the Moon,

To my life give beauty, to my body give strength,

To my corn give goodness, to my house give peace,

To my spirit give truth, to my elders give wisdom.

We must pray for strength.

We must pray to come together.

Pray to the weeping earth

Pray to the trembling waters

And to the wandering rain.

We must pray to the whispering Moon

Pray to the tiptoeing stars

And to the hollering sun.

Taos Pueblo, New Mexico

Hear, Goddess queen, diffusing silver light,
Bull-horn'd, and wand'ring thro' the gloom of Night.
With stars surrounded, and with circuit wide
Night's torch extending, through the heav'ns you ride:
Female and male, with silv'ry rays you shine,
And now full-orb'd, now tending to decline,
Mother of ages, fruit producing moon,
Whose amber orb makes Night's reflected noon:
Lover of horses, splendid queen of night,
All-seeing pow'r, bedeck'd with starry light,
Lover of vigilance, the foe of strife,
In peace rejoicing, and a prudent life:
Fair lamp of Night, its ornament and friend,
Who giv'st to Nature's works their destin'd end.
Queen of the stars, all-wise Diana, hail!
Deck'd with a graceful robe and ample veil.
Come, blessed Goddess, prudent, starry, bright,
Come, moony-lamp, with chaste and splendid light,
Shine on these sacred rites with prosp'rous rays,
And pleas'd accept thy suppliants' mystic praise.

Orpheus: Hymn to the Moon
Onomacritus
6th Century B.C. Greek

Boys and girls, we pledge allegiance
to the moon, virgin Diana,
chastity and innocence,
boys and girls all sing Diana.

O divinity, divinest
fruit of Jove, all powerful sire,
and his Latona, your mother,
gave you birth beneath the sacred.

olive tree of Delos, made you
(sing Diana) mistress of the hills,
young forests, hidden valleys
where winding rivers
disappear in music, sing Diana.

Women in childbirth call upon your name
night goddess, queen of darkness
and false daylight. Sing Diana.

who has steered the circling voyage
of the seasons into years,
bringing with her harvest time
and full granaries and rich farms:

by whatever name we call you,
(sing Diana) hear our prayers,
as years long gone you sheltered us
your sons of Romulus from harm
defend, now and forever, sing Diana!

Dianae Sumus In Fide
Catullus
1st Century B.C. Roman

Calibrating the Cycles

DAY BY DAY

The moon, in its turn, is full and then it's empty. It waxes and then it wanes. Night after night, we watch it get gradually rounder, heavier, fuller. And then it begins to shrink ever so slightly, nightly, slice by crescent slice, until it disappears altogether. Each nocturnal turn of the lunar wheel reveals a subtle alteration in the appearance of the moon.

Tribal cultures bestow upon the days of the moonth wonderfully descriptive names, characteristic of the exact position and phase of the moon. In parts of Polynesia, the string-like first crescent was called "to twist," while the second day was named "crescent." The third and fourth days, when moon-shadows are first visible, are called "the moon has cast a light." The thirteenth day is "the egg" for obvious reasons. The third day after the full moon is "the sea sparkles at the rising." No wonder Gauguin loved it there.

In the East Indies, the eleventh and twelfth day of the lunar cycle, when it is well into its gibbous, or rounded stage, are named "little pig moon" and "big pig moon," respectively. This is not because the moon has gotten porky, but because this is when pigs exhibit an acute agitation in the eerie lunar light of the moon and often escape their pens.

The fourteenth day is "lying," referring to the way the full moon sits on the eastern horizon at sunset. The sixteenth day, "the burner," alludes to the way moonlight comes in through the doors of the houses. The "long tree trunk" and the "short stump," the twenty-sixth and twenty-seventh days of the cycle refer, perhaps, to the increasingly shrinking crescent moon. "Going inside," day twenty-eight is the last visible sliver. On the "inside" days, the moon vanishes altogether.

We in the West, cerebral and abstract as we are, prefer to simply number the days of the cycle of the moon. Ho hum.

Calibrating the Cycles

MOONTH BY MOONTH

People have lived solely by lunar time for most of our conscious existence. For millennia, there was no need to actually count the days in each moon cycle. Although it was determined very early on that the lunar cycle takes twenty-nine and a half days, its exact length was immaterial. What was really important was to know *when* something would happen. For this, it was sufficient to count the moonths.

One moon until the next bleeding. Two moon's walking to reach the pilgrimage site. Six moons until it is likely to rain. Nine moons of growing the baby inside before it is born. Eleven moons of work woven into a rug. Twelve moons until we meet again.

The moonths are frequently associated with and named for common seasonal phenomena: attributes of nature, animal traits or human activities which somehow relate to each particular lunar period. Documents from antiquity mention the "Moon of the Sowing of the Rice" in China; the "Moon of the Garlic Harvest" in Incan Peru and the "Moon of the Reaping of the Harvest" in ancient Persia.

The Mayan calendar had twenty moonths. Many of the glyphs which stand for these have been translated: Mat Month, Frog Month, Goddess Month, Bat Month, Summer Month, Green Month, White Month, Deer Month, Ribs Month, Falcon Month and Turtle Month. Cultures from various regions of Native North America boasted Falling Leaf Moon, War Moon, Sore Eye Moon, Hunger Moon, Moon of Popping Trees, Silver Salmon Moon and so on.

How the moonths are named says a lot about a people, revealing specific details of environment, weather conditions, seasonal occupations, diet and belief systems. The twelve moonth names of the Omaha, dwellers of the Great Plains and woodlands of the Missouri

River valley in what is now Nebraska, clearly indicate that they were hunters, focused as they are primarily on animals:

MOON IN WHICH THE SNOW DRIFTS INTO
THE TENTS OF THE HOGA
MOON IN WHICH THE GEESE COME HOME
LITTLE FROG MOON
MOON IN WHICH NOTHING HAPPENS
MOON IN WHICH THEY PLANT
MOON IN WHICH THE BUFFALO
BULLS HUNT THE COWS
MOON IN WHICH THE BUFFALO BELLOW
MOON IN WHICH THE ELK BELLOW
MOON IN WHICH THE DEER PAW THE EARTH
MOON IN WHICH THE DEER RUT
MOON IN WHICH THE DEER SHED THEIR ANTLERS
MOON IN WHICH THE BLACK BEARS ARE BORN.

The calendar of their neighbors about five hundred miles to the north, the Ojibwa, reflects a completely different lifestyle. Here, along the heavily forested waterways surrounding the western Great Lakes, agriculture was impractical and large prey scarce. The people thrived on the wild fruits and grains that they gathered:

Long Moon
Spirit Moon
Moon of the Suckers
Moon of the Crust on the Snow
Moon of the Breaking of Snowshoes
Moon of the Flowers and Blooms
Moon of Strawberries
Moon of Raspberries
Moon of Gathering Wild Rice
Moon of the Falling Leaves
Moon of Freezing
Little Moon of the Spirit.

The Ugric Ostiak, a group living further north still on the vast, empty tundra of northern Siberia has produced moon names which reflect their chilly existence. Trees seem to have been prized for their rarity and the importance of their wood — less for fuel than for shelter for themselves and their horses. The list also suggests that fish and gamebirds are important food staples:

Spawning Month
Pine-Sapwood Month
Birch-Sapwood Month
Salmon-Weir Month
Month of Hay Harvest
Ducks and Geese Go Away Month
Naked Tree Month, Pedestrian Month
Month of Going Home While Ice Still Remains
Month of Going on Horseback
Great Month
Little Winter-Ridge Month
Windy Month of Crows.

The Chinese calendar alternates "big months" of thirty days and "small months" of twenty-nine days. The moonths, themselves, are then subdivided into mini-moonths which correspond to the waxing and waning phases of the moon. Each new and full moon designates the beginning of a specified term, twenty-four in all:

The Spring Begins
The Rain Water
The Excited Insects
The Vernal Equinox
The Clear and Bright
The Grain Rains
The Summer Begins
The Grain Fills
The Grain in Ear
The Summer Solstice
The Slight Heat
The Great Heat
The Autumn Begins
The Limit of the Heat
The White Dew
The Autumn Equinox
The Cold Dew
The Hoar Frost Descends
The Winter Begins
The Little Snow
The Heavy Snow
The Winter Solstice
The Little Cold
The Severe Cold

The Jewish calendar, too, reflects the prevailing conditions of each monthly period. The Hebrew moonths are named from the old

Babylonian moonths, which, in turn, received their names from the Assyrian. *Tishri*, the first moonth, comes from the Syrian, *shera*, or *sherei,* which means, "to begin." *Tevet*, which falls around December, normally a wet and muddy time, takes its name from *tava*, "to sink in." *Iyyar*, from the Hebrew word, *or*, meaning, "light," is the moonth in which the Vernal Equinox occurs and the days grow longer.

Early Arabic calendars once aligned the moonths with the solar year. But the current Islamic calendar, adopted in the 7th Century A.D., is purely lunar. Since the moonths travel through the entire year, the names have lost their original seasonal correspondence. *Safar*, the second moonth comes from the Arabic, *safara*, "to become empty" referring to the granaries at the end of the time of plenty. *Rabi*, the name of both the third and fourth moonths, describes the time when the earth becomes green after the autumn rains. *Jumada*, the fifth and sixth moonths, meaning "to become hard" or "to freeze," must have once been in winter. *Ramadan*, the ninth moonth, comes from *ramada*, "to be heated by the sun."

The names of the old Dutch moonths are remarkably similar to those used by the Anglo-Saxons a thousand years ago.

JANUARY	Lauwmaand	*Chilly Month*
FEBRUARY	Sprokelmaand	*Vegetation Month*
MARCH	Lentmaand	*Spring Month*
APRIL	Grasmaand	*Grass Month*
MAY	Blowmaand	*Flower Month*
JUNE	Zomermaand	*Summer Month*
JULY	Hooymaand	*Hay Month*
AUGUST	Oostmaand	*Harvest Month*
SEPTEMBER	Herstmaand	*Autumn Month*
OCTOBER	Wynmaand	*Wine Month*
NOVEMBER	Slagtmaand	*Slaughter Month*
DECEMBER	Wintermaand	*Winter Month*

The moonths of the Druids, animistic worshippers of tree spirits, were named in honor of Birch, Rowan, Ash, Alder, Willow, Hawthorne, Oak, Holly, Hazel, Vine, Ivy, Reed and Elder.

Several cultures over the ages have devised flower calendars with every moonth designated by a characteristic flower. These were not business or religious time-keeping calendars, but, rather, associative, illustrative tools used for artistic and allegorical purposes, poetry and painting.

	JAPANESE	CHINESE	VICTORIAN
January	*Pine*	*Plum Blossom*	*Snowdrop*
February	*Plum*	*Peach Blossom*	*Primrose*
March	*Peach*	*Tree Peony*	*Violet*
April	*Cherry*	*Cherry Blossom*	*Sweet Pea and Daisy*
May	*Iris*	*Magnolia*	*Lily of the Valley*
June	*Wisteria*	*Pomegranate*	*Rose*
July	*Morning Glory*	*Lotus*	*Water Lily*
August	*Lotus*	*Pear Blossom*	*Poppy and Gladiolus*
September	*Chrysanthemum*	*Mallow*	*Morning Glory*
October	*Autumn Nanakusa*	*Chrysanthemum*	*Calendula*
November	*Maple*	*Gardenia*	*Chrysanthemum*
December	*Bamboo*	*Poppy*	*Flowerless*

Keeping their priorities straight, the French Revolutionary ten-month calendar (which was arbitrary and not moon related) named the first month, *Venemaire*, Fine Wine Month; followed by: *Brumaire*, Fog Month; *Frimaire*, Frost Month; *Nivose*, Snow Month; *Pluviose*, Rain month; *Ventose*, Wind Month; *Germinal*, Seed Month; *Floreal*,

Flower Month; *Prairial*, Meadow Month; *Messidore*, Harvest Month; *Thermidor*, Heat Month; and *Fructidor*, Fruit Month.

Compared with these, our month names fall flat — tame and boring. We have inherited our months intact from the Roman calendar, instituted by Julius Caesar in 45 B.C. The names signify absolutely nothing to us anymore, and are, as far as most of us know, practically devoid of meaningful allusions to the natural world.

Januarius, the New Year month was named for the god, Janus, who looks both backward and forward in time; *Februarius* was for Februus, the god who oversees the cleansing of sins; *Martius* was for the war god, Mars, perhaps in deference to March's stormy weather; *Aprilis*, from the Latin, *aperire*, means "to open" or "to bud;" *Maius* was in honor of Maia, Goddess of Green Growth; *Junius*, from the Latin, *junores*, "young people," might refer to the fertility festivals celebrated around the Summer Solstice.

Julius was named for Julius Caesar, author of the calendar and *Augustus* was for Augustus, Caesar's grandnephew and heir. As if the cup of inspiration had run dry after allocating the eighth name, the remaining months were given numbers, which, having once belonged to a previous and outdated calendar, aren't even correct. The ninth, tenth, eleventh and twelth months, September, October, November and December, actually mean "seven," "eight," "nine" and "ten."

Calibrating the Cycles

LUNAR CALENDARS

))))●●●●●((((

THE MOON, DOUBTLESS, IS THE YEAR, AND ALL LIVING BEINGS.

Satapatha-Brahmana,
Ancient Hindu Vedic Commentary

THE MOON WAS CREATED FOR THE COUNTING OF THE DAYS.

Hebrew Midrash Text

THE MOON AND ITS PHASES GAVE MAN HIS FIRST CALENDAR.

Isaac Asimov
20th Century American

The movement of the moon measures our days and divides our years. From earliest times, our ancestors learned to keep track of its endlessly repeated path across the sky, so that they might keep up — not with the Joneses, mind you, but with the cycles of nature, herself. It was crucial to survival to stay in tune with the changes in the sky, in the weather, in the world around them.

The moon has, from our earliest beginnings, defined time for us. Starkly silver as it is against the celestial abyss, the moon is the only luminary in the sky whose change of position produces a shape-shift visible to the naked eye. This makes it eminently easy to observe and chart.

A 30,000 year-old piece of bone found in the Dordogne region of France is inscribed with a series of notches that seem to have been made by different instruments over a certain amount of time. These notations are thought to indicate the lunar cycle. Two similar bones, both about 8,500 years old, were also unearthed — one in equatorial Africa and one in Czechoslovakia. They are both incised with repeating sets of fifteen and sixteen notches in precisely ordered intervals.

The so-called Ishango bone found at the source of the Nile can be read as a basic lunar calendar. It plots the pattern of time that transpires between the first sighting of the new crescent moon and the night when the moon attains its fullest circle. This prehistoric record is accurate for a five-and-a-half month period.

The counting of each moonth starts with the first glimpse of the slimmest lunar visage, the inaugural instant of re-visibility which is clearly discernible in the dark sky. It ends when the moon disappears once again. In 631 A.D., Mohammed declared that the Islamic calendar should have twelve lunar moonths, and that the moonth shall begin when two faithful Muslims together have observed the first crescent from the top of a mountain or an open plain. Prominent in Arabic art for thousands of years, the crescent moon motif, the *hilal*, has graced the standard of Islam for two hundred years.

Corroboration of the first viewing was also evidently important in Mesopotamia. A letter to the Assyrian king, Esarhaddon, sent sometime between 680 B.C. and 669 B.C. recounts:

On the thirtieth I saw the moon. It was in a high position. The king should wait for the report from the city of Assur and then may determine the first day of the month.

Early Hebrews employed fire signals and, later, messengers, to convey to the entire community the lawful sighting of the new moon and thus the beginning of the associated lunar festivals. Ancient Greek criers would loudly announce the rising of the new moon, the beginning of the new moonth.

In Rome, a pontifex minor would take watch for the first moon from the top of the Capitoline Hill. Upon seeing it, he would call out to Juno, the Queen of the Gods. The first day of each Roman moonth was called the *calends* which means, "to call out." The English words "moon," "month," and "measure" were spawned from the same Indo-European root. In the Korean language, too, the word for "moon" and "month" are identical.

A lunar calendar which provides a general sense of passing time is completely adequate for the needs of those who wander. Though nomads often travel in the cool safety of the protective night guided by the light of the moon, they actually live from day to day, gathering and hunting provisions as they go.

But those who stay in one place and grow their food become much more directly dependent upon the sun and its annual cycles reflected in the seasonal changes. It would behoove farming peoples to plot the sun's course in order to be able to correctly determine the best times to plant, tend and harvest their crops. With the development of agriculture and a sedentary way of life, came the perceived need for and the subsequent invention of a solar calendar.

Over the past few thousand years, the awkward solar calendar has undergone many alterations and refinements, one of which was the invention of the month as an arbitrary, arithmetic division of the year. Calendar months of 30 and 31 days are longer than the actual period between one full moon and the next which is 29.53 days. The months as they have become delineated have thus lost their intrinsic correlation with the actual phases of the moon. And as a result, humankind has surrendered our once-intimate relationship with our loyal lunar satellite.

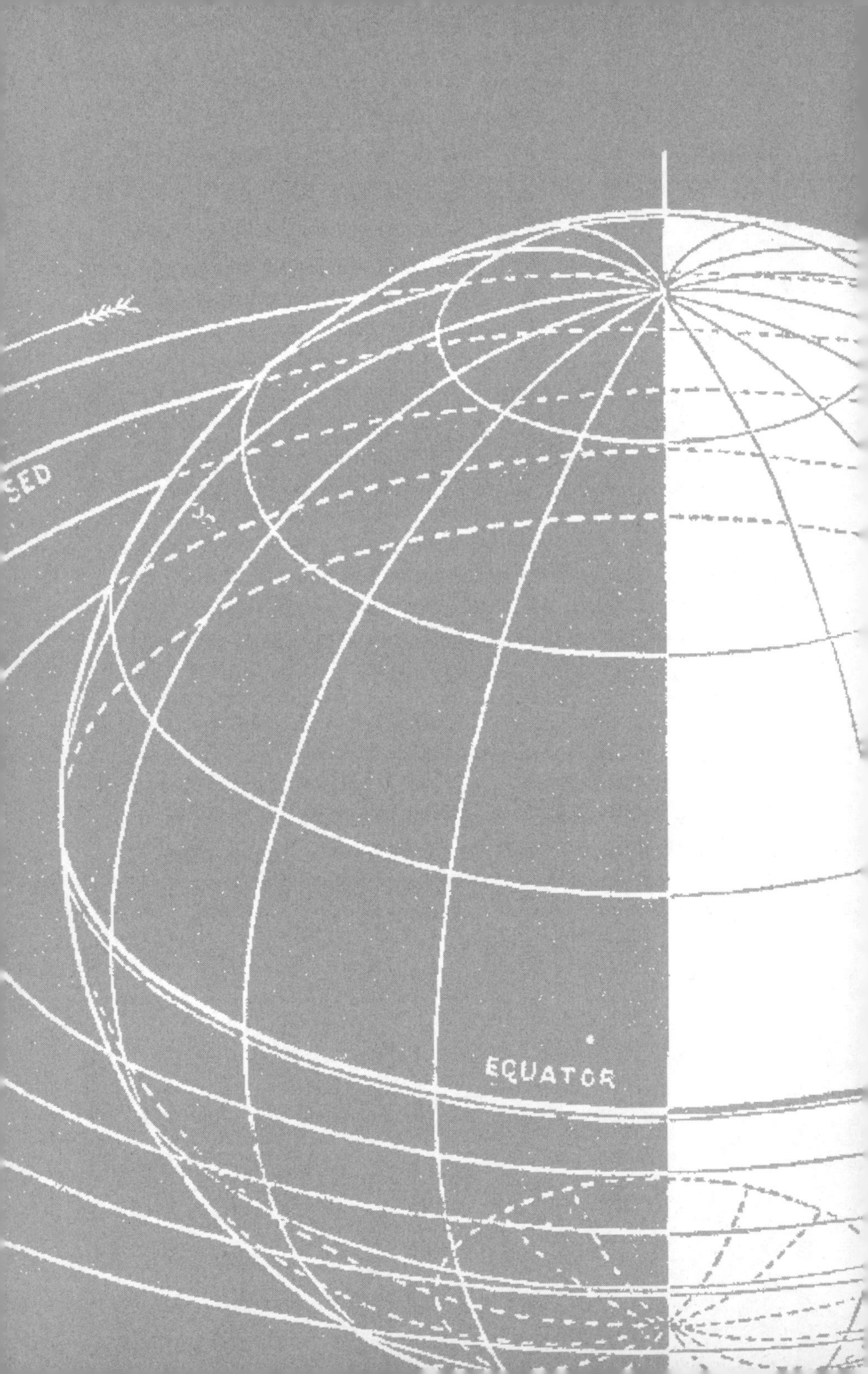
EQUATOR

Facts About the Moon

What We Know

Origins

Speculations on the Origins of the Moon

))))●●●●((((

With the coming of the Scientific Age, mythical descriptions of the nature and origin of the moon began to be replaced with theories based on technologically enhanced observation and measurement. Central to this revolution in astronomy was the refinement of the telescope by Galileo in the 17th Century.

From 1796 to about 1900, it was widely accepted by scientists that the moon as well as the planets were formed by gaseous clouds which condensed as they rotated and spun through space. This Nebular Hypothesis was eventually disproved and replaced by three other theories:

- The moon began as a small rotating body that was already circling around the earth. This lunar seedling grew into its present size gradually, gaining girth from the space debris that it accumulated and assimilated along its course.

- The moon was once part of the earth. At some point, the earth began spinning too quickly and became unstable. The moon was formed when solar tidal forces ripped out a chunk of the parent planet at the equator and sent it sailing into orbit.

- The moon was formed somewhere else in the universe and pulled by powerful gravitational attraction into its orbit around the earth.

In 1967, the premiere unmanned space probe landed on the moon and collected samples of lunar soil and rock, facilitating for the first time the opportunity for hands-on observation and study. None of the prevailing speculations were supported by science's extensive analysis of the substance of the moon. A new explanation was then formulated.

According to current thought, the moon was formed by a cataclysmic collision between the earth and a Mars-sized planet when the earth was still in a molten state. The impact stripped the earth of its mantle, exposing its iron core and initiating a blast of silicate gases into space. This immense cloud of hot gas and debris surrounded the earth at first, and then it cooled and coalesced into the moon. Once, in the beginning, the orbit of the moon carried it closer to Earth. Over thousands of years, the moon has moved further away, pulling loose space particles into itself as it went.

Natural Attributes

Vital Statistics

Diameter
2,160 miles
0.27 of Earth's diameter 7,910 miles

Circumference
6,790 miles

Mean Radius
1,080 miles

Mass
$8x10^{19}$ tons
0.0123 Earth's mass
1/81 that of Earth
$3.7x10^{-8}$ of the sun's mass

Weight
81 quintillion tons
Earth weighs 588 quintillion tons

Volume
0.0204 Earth's volume

Density
3.34 times more dense than water

Surface Area
15 million square miles
1/14 that of Earth

Surface Gravity
5.31 $feet^2$ per second
0.165 16% times Earth's gravity

Surface Temperature
273 degrees Fahrenheit day time
-280 degrees Fahrenheit night time

Visible Surface
59% at any one time

Mean Orbital Speed
2,287 miles per hour

Orbital Direction
Counterclockwise

Apogee: Greatest Distance from Earth
252,710 miles

Perigee: Shortest Distance from Earth
221,463 miles

Mean Distance from Earth
238,657 miles

Average Albedo
(Percent of sunlight reflected) 7
Earth's albedo is 40

Rotation Period
27 days, 7 hours, 43 minutes, 11.5 seconds

Revolution of Perigee Period
3,232 days

Anatomy

Atmosphere	*Practically None*
Water	*None*
Outer Layer of Crust	*36–62 miles thick*
Soil Cover	*2–3 feet deep*
Surface Dust	*1/8 inch–3 inches thick*
Organic Composition of Soil	*None*

Inorganic Composition of Soil

Lunar soil is composed mostly of surface rock that has been broken up over billions of years of meteor bombardment. Lunar rock, which is slightly magnetic, was found to contain no unknown substances or chemical elements. 42% of the atomic structure of the moon is oxygen. Aluminum, nitrogen, calcium, titanium, magnesium, and ytterbium all appear more abundantly than on Earth.

(IN) PRELIMINARY CONCLUSION, THE MATERIAL FROM ALL THREE SEAS — THE SEA OF TRANQUILLITY, THE OCEAN OF STORMS AND THE SEA OF FERTILITY — IS SURPRISINGLY SIMILAR IN ITS PETROLOGICAL, MINERALOGICAL AND CHEMICAL COMPOSITION, THOUGH CERTAIN DETAILS ARE DIFFERENT."

Aleksandr P. Vinogradov
20th Century Russian Geochemist

Lunar Litter

50 tons were left by astronauts including a flag, several golf balls and tire tracks.

Moon Rocks

The moon rock samplings brought back by Apollo astronauts cost roughly $3 million dollars per ounce if divided by the program's cost.

IT (A MOON ROCK) LOOKS LIKE A DIRTY POTATO. I'M AFRAID THEY ARE GOING TO HAVE TO CLEAN UP THE MOON BEFORE I GO THERE.

Margaret Mitchell
Wife of Attorney General John Mitchell

Surface Topography

Though dead and sterile, the surface of the moon is in a constant state of slow change due to moonquakes and ever-present meteor strikes. The moon is so susceptible to meteorites because it has no atmosphere to burn them up, the way Earth does. Many of the dramatic surface details of the moon are visible with the naked eye. The details are even more astonishing with binoculars or a modest amateur telescope.

Two early American astronauts described the moon from their vantage point aboard the Apollo 14 spacecraft while in lunar orbit in 1971:

IT LOOKS LIKE IT'S BEEN MOLDED OUT OF PLASTER OF PARIS.

Commander Edgar D. Mitchell, Jr.

HEY, YOU'RE NOT GOING TO BELIEVE THIS. IT LOOKS JUST LIKE THE MAP.

Major Stuart A. Roosa

Natural Attributes

Surface Features

I FEEL SURE THAT THE SURFACE OF THE MOON
IS NOT PERFECTLY SMOOTH, FREE FROM
INEQUALITIES, AND EXACTLY SPHERICAL,
AS A LARGE SCHOOL OF PHILOSOPHERS
CONSIDERS WITH REGARD TO THE MOON
AND THE OTHER HEAVENLY BODIES, BUT THAT,
ON THE CONTRARY, IT IS FULL OF INEQUALITIES,
UNEVEN, FULL OF HOLLOWS AND PROTUBERANCES,
JUST LIKE THE SURFACE OF THE EARTH ITSELF,
WHICH IS VARIED EVERYWHERE BY LOFTY
MOUNTAINS AND DEEP VALLEYS

Galileo Galilei
17th Century Italian astronomer

Craters

The most striking lunar objects seen through a telescope, they were likely formed by meteor impact. Dimensions range from the almost invisible to enormous ring plains, the largest of which has a diameter of 180 miles. There are at least 30,000 craters whose diameter is in excess of 1 km. Some have flat, smooth floors, others are filled with large complexes of mountains. The average height of a lunar crater wall is 9,000-18,000 feet.

Marias

These so-called lunar "seas" appear as large gray patches and are visible to the naked eye. They are vast, arid, circular plains sunk below the general surface level of the moon. They never contained water, but are actually huge lava flows which have solidified into basalt.

Mountains

Appearing as ranges, peaks, uplands and isolated hills, lunar mountains cover as much surface area as the marias. They commonly constitute the walls of the ring-like crater formations. The Leibbnitz Mountains at the south lunar pole are the highest range at 30,000 feet in altitude.

Paludes

These are "marshes" which are similar to the marias, but smaller in area.

Rills

Immense trench-like cracks on the surface, rills run parallel to the mountains. The longest is 184 miles in length. There are more than 2,000 known rills.

Sinii

Like the "seas" and "marshes," these "bays" and "gulfs" are also dry.

The Ten Largest Lunar Craters

Crater	Diameter
Bailly	*184 miles*
Clavius	*140 miles*
Schickard	*139 miles*
Grimaldi	*138 miles*
Humboldt	*130 miles*
Schiller	*112 miles*
Petavius	*110 miles*
Maginus	*101 miles*
Riccioli	*94 miles*
Hipparchus	*93 miles*

Lunar Seas

Latin Name	English Name
Sinus Aestuum	Bay of Heats
Mare Australe	Southern Sea
Mare Crisium	Sea of Crises
Palus Epidemiarum	Marsh of Epidemics
Mare Fecunditatis	Sea of Fertility
Mare Frigoris	Sea of Cold
Mare Humboldtinaum	Humboldt's Sea
Mare Humorum	Sea of Humours
Mare Imbrium	Sea of Showers
Sinus Iridum	Bay of Rainbows
Mare Marginis	Marginal Sea
Sinus Medii	Central Bay
Lacus Mortis	Lake of Death
Palus Nebularum	Marsh of Mists
Mare Nectaris	Sea of Nectar
Mare Nubium	Sea of Clouds
Mare Orientale	Eastern Sea
Oceanus Procellarum	Ocean of Storms
Palus Putredinis	Marsh of Decay
Sinus Roris	Bay of Dews
Mare Serenitatis	Sea of Serenity
Mare Smythii	Smyth's Sea
Palus Somnii	Marsh of Sleep
Lacus Somniorum	Lake of Dreamers
Mare Spumans	Foaming Sea
Mare Tranquillitatis	Sea of Tranquillity
Mare Undarium	Sea of Waves

Natural Attributes

Monthly Cycles

A month is understood to be the time that it takes the moon to travel 360 degrees around the earth. While the moon is revolving around the earth, the earth, too, is moving forward in its own orbit around the sun. Consequently, the moon must travel an additional 53 hours before it regains its original relative position to both the earth and the sun.

This point-of-return full-circle, so to speak, can be counted from any number of reference points. And because the moon's orbit is elliptical rather than round, these points will diverge slightly in their timing.

Synodic or Lunar Month

Moon To Moon

The period from one new or full moon to the next.

29.53059 Days

29 Days, 12 Hours, 44 Minutes, 2 Seconds

Sidereal Month

Star To Star

The time it takes for the moon to complete one revolution around the earth and return to its starting position, using a fixed star as a reference point.

27.32166 Days

27 Days, 7 Hours, 43 Minutes, 11.5 Seconds

Anomalistic Month

Perigee To Perigee or Apogee To Apogee

The amount of time it takes the moon to return to the point in its orbit when it is either closest or furthest from the earth.

27.55455 Days

27 Days, 13 Hours, 18 Minutes, 33.2 Seconds

Nodical or Draconic Month

Node to Node

The period of time it takes the moon to return to the same point in its orbit where it intersects the ecliptic, the apparent annual path of the sun around the celestial sphere.

27.21222 Days

27 Days, 5 Hours, 5 Minutes, 35.8 Seconds

Tropical Month

First Point of Aries To First Point of Aries

The length of time it takes the moon to travel through the twelve constellations of the zodiac and return to the beginning.

27.321582 Days

27 Days, 7 Hours, 43 Minutes, 4.7 Seconds

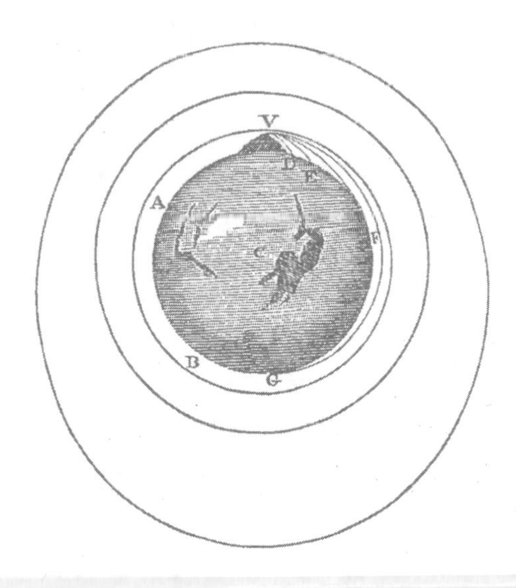

Explorations

Illustrious Moon Watchers

IT IS A MOST BEAUTIFUL AND DELIGHTFUL
SIGHT TO BEHOLD THE
BODY OF THE MOON

GalileoGalilei
17th Century Italian astronomer

Nicolas Copernicus

(1473-1543) Polish

The first to theorize that the earth and other planets revolved around the sun and that the moon revolved around the earth.

Tycho Brahe

(1546-1601) Danish

Though he believed in an earth-centered universe, he correctly measured the lunar orbit including the slight variations caused by the sun's gravity.

Galileo Galilei

(1564-1642) Italian

He created and refined telescopic observation of the moon, which allowed for the first detailed lunar maps.

Johannes Kepler

(1571-1630) German

A believer in Copernicus' theory of a sun-centered universe, he made relevant observations of the orbit of the moon, its effects on the tides, and the nature of the lunar surface.

Giovanni Riccioli

(1598-1671) Italian

Many of the names he gave to the features and famous landmarks of the lunar surface are still in use today.

Johannes Hevelius

(1611-1687) Polish

A specialist in the study of the lunar surface, he published a detailed description of the moon. He discovered that the lunar seas or marias were actually great plains.

Jean Dominique Cassini

(1625-1712) Italian-born French

The director of the Paris Observatory, he mapped the moon.

Johann Mayer

(1723-1762) German

He created the first map of the moon which used a system of coordinates.

Johann Hieronymous Schroeter

(1745-1816) German

The founder of selenography, the modern science of moon study, he devoted most of his life to producing moon drawings. His published studies of lunar landscapes were the first to include measurements.

Sir John Herschel

(1792-1871) English

The "father of modern astronomy," he discovered Uranus and thought that the moon was habitable. His discoveries about the moon were sensationalized by the popular press including The New York Times and included phony reports of strange lunar animals.

John Draper

(1811-1882) American

He took the first photograph of the moon on March 23, 1840.

Lewis Rutherford

(1816-1892) American

A lawyer by profession, he was one of the pioneers of lunar photography.

Edmund Neison

(1851-1938) English

His book *The Moon* was published when he was only 25. It contained a 2 foot lunar map that described every surface formation then known.

William Henry Pickering

(1858-1938) American

He made many significant observations and established several major observatories.

Bernard Lyot

(1897-1952) French

Considered to be one of the world's most important astronomers, he accomplished important research about the surface of the moon.

Explorations

Walking on the Moon

OF ALL THE CREATURES WHO HAD YET WALKED ON EARTH, THE MAN-APES WERE THE FIRST TO LOOK STEADFASTLY AT THE MOON. AND THOUGH HE COULD NOT REMEMBER IT, WHEN HE WAS VERY YOUNG MOON-WATCHER WOULD SOMETIMES REACH OUT AND TRY TO TOUCH THAT GHOSTLY FACE RISING ABOVE THE HILLS. HE HAD NEVER SUCCEEDED, AND NOW HE WAS OLD ENOUGH TO UNDERSTAND WHY, FOR FIRST, OF COURSE, HE MUST FIND A HIGH ENOUGH TREE TO CLIMB.

2001: A Space Odyssey
Arthur C. Clarke, 20th Century English

Perhaps the first historical description of a fictional voyage to the moon was created by Lucian of Samos in 165 A.D. He envisioned an expedition of lunar explorers from Earth who come across a huge mirror hung over a shallow pool. In it they could see reflected all of Earth in great detail — every nation, every city.

Lucian's fantasy was realized 1803 years later by the Apollo 8 astronauts on humankind's first manned orbit around the moon. They, too, were able to peer down upon their home planet from space, to perceive it as a unified whole and in perspective with the vast universe. One year later, in 1969, Apollo 11 landed on the moon's surface at the Sea of Tranquillity. Aboard were the first earthlings to set foot on lunar soil.

And so, the Man in the Moon, the Mother in the Moon, the Maid in the Moon, and all the sundry lunar pets have now, in this era of mass media consciousness, been replaced and far surpassed by that vastly impressive, permanently indelible collective image that we all share — the Man Walking on the Moon.

Explorations

Lunar Missions

Unmanned Flybys, Orbits, and Landings

Date	Spacecraft	Mission
Oct 1958	Pioneer 1 (USA)	*Flyby*
Dec 1958	Pioneer 3 (USA)	*Flyby*
Jan 1959	Luna 1 (USSR)	*Flyby*
Mar 1959	Pioneer 4 (USA)	*Passed within 37,300 miles of the moon*
Sep 1959	Luna 2 (USSR)	*First lunar impact east of Mare Serenitatis area*
Oct 1959	Luna 3 (USSR)	*First photographs of the lunar far side (dark side)*
Apr 1962	Ranger 4 (USA)	*Crashed on lunar far side*
Feb 1964	Ranger 6 (USA)	*Cameras failed/impacted in Mare Tranquillitatis area*
Jul 1964	Ranger 7 (USA)	*Transmitted first close-up photos of the moon/impacted in the Mare Nubium area*
Feb 1965	Ranger 8 (USA)	*Transmitted first high-quality photos of the moon/impacted in Mare Tranquillitatis area*
Mar 1965	Ranger 9 (USA)	*Transmitted photos/impacted in the Alphonsus Crater*
May 1965	Luna 5 (USSR)	*First soft landing attempt/crashed in the Mare Nubium area*
Jul 1965	Zond 3 (USSR)	*Photographed far side*

Date	Spacecraft	Mission
Oct 1965	Luna 7 (USSR)	*Soft landing failed/ crashed in the Oceanus Procellarum area*
Dec 1965	Luna 8 (USSR)	*Soft landing failed/ crashed in the Oceanus Procellarum area*
Jan 1966	Luna 9 (USSR)	*First successful soft landing (in the Oceanus Procellarum area)/first television transmission from the lunar surface*
Mar 1966	Luna 10 (USSR)	*First lunar satellite/studied lunar surface radiation, magnetic field intensity and gravitation*
May 1966	Surveyor 1 (USA)	*First American soft-landing (in the Oceanus Procellarum area)/robotic laboratory*
Aug 1966	Lunar Orbiter 1 (USA)	*Photographed over 2 million square miles of the moon's surface/impacted on the far side*
Aug 1966	Luna 11 (USSR)	*Orbited*
Sep 1966	Surveyor 2 (USA)	*Soft landing failed/crashed near the Crater Copernicus*
Oct 1966	Luna 12 (USSR)	*Orbited/transmitted large-scale photographs*
Nov 1966	Lunar Orbiter 2 (USA)	*Orbited/photographed landing sites/impacted on 10/11/67*
Dec 1966	Luna 13 (USSR)	*Soft-landed at Oceanus Procellarum/ measured soil density and surface radioactivity*

Date	Spacecraft	Mission
Feb 1967	Lunar Orbiter 3 (USA)	*Orbited/sent photos and data/impacted 10/9/67*
Apr 1967	Surveyor 3 (USA)	*Soft-landed (at Oceanus Procellarum)/robotic laboratory/ran soil tests*
May 1967	Lunar Orbiter 4 (USA)	*Orbited/took first pictures of the lunar south pole/ impacted 10/6/67*
Jul 1967	Surveyor 4 (USA)	*Radio contact with spacecraft lost 2 1/2 minutes prior to landing in the Sinus Medii area*
Jul 1967	Explorer 35 (USA)	*Orbited/measured magnetic fields*
Aug 1967	Lunar Orbiter 5 (USA)	*Orbited/impacted 1/31/68*
Sep 1967	Surveyor 5 (USA)	*Soft-landed (in Mare Tranquillitatus area)/ robotic laboratory/ conducted soil tests*
Nov 1967	Surveyor 6 (USA)	*Soft-landed (in the Sinus Medii area)/robotic laboratory*
Jan 1968	Surveyor 7 (USA)	*Soft-landed (at Tycho Crater)/robotic laboratory*
Apr 1968	Luna 14 (USSR)	*Orbited/studied gravitational field*
Sep 1968	Zond 5 (USSR)	*First flyby and Earth return*
Nov 1968	Zond 6 (USSR)	*Lunar flyby and Earth return*
Jul 1969	Luna 15 (USSR)	*Sample returner/crashed in the Mare Crisium area*
Aug 1969	Zond 7 (USSR)	*Lunar flyby and Earth return*

Date	Spacecraft	Mission
Sep 1970	Luna 16 (USSR)	*First landing and Earth return/robotic soil sample (from the Mare Foecunditatis area)/returned to Earth*
Nov 1970	Zond 8 (USSR)	*Lunar flyby and Earth return*
Nov 1970	Luna 17 (USSR)	*First robotic rover landed in the Sinus Iridium/Earth return*
Sep 1971	Luna 18 (USSR)	*Crashed in the Mare Foecunditatis area*
Sep 1971	Luna 19 (USSR)	*Orbited/studied lunar gravitational field*
Feb 1972	Luna 20 (USSR)	*Landed at Mare Crisium basin rim/collected and returned soil samples*
Jan 1973	Luna 21 (USSR)	*Landed at Mare fill of Le Monnier/robotic lunar rover brought home soil samples*
May 1974	Luna 22 (USSR)	*Orbited/returned soil samples*
Nov 1974	Luna 23 (USSR)	*Landed at Mare Crisium/ damaged on landing*
Aug 1976	Luna 24 (USSR)	*Landed at Mare Crisium/ collected and returned soil samples*
Jan 1990	Hiten (Muses-A) (Japan)	*Orbited/studied gravity's effect on satellites/crashed on the moon 4/11/93*
Jan 1994	Clementine (USA)	*Orbited/photographed the lunar surface*
Jan 1998	Lunar Prospector (USA)	*Orbited/mapped elemental composition of lunar crust/produced no evidence of water/controlled crash into a lunar crater 7/31/99*

Manned Orbits, Landings, and Walks
Apollo (USA) Program

Date	Spacecraft/Crew	Mission
Oct 11, 1968	**Apollo 7** Cunningham, Eisele, Shirra	*Test orbits around earth*
Dec 21, 1968	**Apollo 8** Anders, Borman, Lovell	*First manned moon orbit*
Mar 3, 1969	**Apollo 9** McDivit, Scott, Schweickart	*Test orbits around earth*
May 18, 1969	**Apollo 10** Cernan, Stafford, Young	*Docking maneuvers in lunar orbit, tested for landing*
Jul 16, 1969	**Apollo 11** Aldrin, Armstrong, Collins	*First humans on the moon/ landed at Mare Tranquillitatis (7/20/69)*
Nov 12, 1969	**Apollo 12** Bean, Conrad, Gordon	*Landed at Oceanus Procellarum (11/19/69)*
Apr 11, 1970	**Apollo 13** Haise, Lovell, Swigert	*Aborted landing attempt after spacecraft explosion*
Jan 31, 1971	**Apollo 14** Shepard, Mitchell, Roosa	*Landed at Fra Mauro (2/5/71)*
Jul 26, 1971	**Apollo 15** Scott, Irwin, Worden	*Landed in Apennius-Hadley region/ used Lunar Rover to explore surface (7/30/71)*
Apr 16, 1972	**Apollo 16** Duke, Mattingly, Young	*Landed at Descartes Highlands (4/20/72)*
Dec 7, 1972	**Apollo 17** Cernan, Evans, Schmitt	*Landed at Taurus-Littrow area/first geologist on the moon (12/11/72)*

WHO (SAID THE MOON)
DO YOU THINK I AM AND
PRECISELY WHO
PIPSQUEAK, WHO ARE YOU

WITH YOUR UNCIVIL LIBERTIES
TO DO AS YOU DAMN PLEASE?
BOO!

I AM THE SERENE
MOON (SAID THE MOON).
DON'T TOUCH ME AGAIN.

TO YOUR POKING TELESCOPES,
YOUR PEEKING EYES
I HAVE LONG BEEN WISE.

SCIENCE? ANOTHER WORD
FOR MONKEYSHINE.
YOU HEARD ME.

GET DOWN, LITTLE MAN, GO HOME,
BACK WHERE YOU COME FROM,
BAH!

OR MY GOLD WILL BE TURNING GREEN
ON ME (SAID THE MOON)
IF YOU KNOW WHAT I MEAN.

"Edith Sitwell Assumes
the Role of Luna"
Robert Francis
20th Century American

Speaking of the Moon

Encyclopedia of Lunar Terminology

Honeymoon: It was the custom in northern Europe for the bride and groom to drink honey mead every day for 30 days after their wedding. Honey symbolizes the bee's qualities as a fertile pollinator.

Loony bin: A hospital or sanitarium for the mentally ill. From the moon, Luna.

Lucent: Giving forth light. Bright.

Lucid: Thinking clearly. Expressing meaning with clearness without ambiguity.

Lucifugous: Avoiding, turning away from the light.

Lucifer: Biblical Satan. A negative version of the ancient myths which refer to the moon as the "Horned One" and as the home of dead souls. 2. Light, light bearing.

Lunacy: An intermittent form of insanity supposed to vary in intensity with the phases of the moon.

Lunar: Pertaining to or determined by the moon.

Lunarian: Inhabitant of the moon or one who possess special knowledge of it.

Lunate: Crescent shaped.

Lunatic, Maniac: One mentally deranged by a condition originally thought to be caused by possession by the spirit of the moon, known in old Europe as both Luna and Mana.

Lunation: Period of time between new moons.

Lune: Figure shaped like a half moon or crescent.

Lunette: From the French, "little moon." Applied to a number of objects having crescent shapes, but especially refers to eye-glasses.

Luniolatry: Worship of the moon.

Mania: Violent madness, form of insanity characterized by excessive excitement or enthusiasm.

Menopause: Period of life when menstruation ceases (and presumably the monthly influence of the moon).

Mensa: Society for those with especially high I.Q.'s. From the Roman moon goddess Mensa, ruler of measurement, calendars, numerical systems.

Menses: Monthly discharge of blood from the uterus which tends to occur in sync with the moon's phases. Derived from the Greek *mene*, "month" and "moon."

Menstruation: "Moon change." The cyclical process of releasing the menses. In Germany, it is referred to as *die monde*, "the moon;" in France, it is called *le moment de la lune*, "the time of the moon."

Menstruum: Alchemical term for a solution that dissolves a solid body. A solvent.

Mensurable: Capable of being measured. 2. Having a fixed time or rhythm.

Moonbeam: A ray or beam of light from the moon.

Moonblind, Moonblink: A condition of temporary night blindness thought to be brought on by sleeping in moonlight, especially in tropical climates. It is now recognized as a symptom of vitamin A deficiency.

Moonbow: A prismatic display less intense than a rainbow produced by moonlight refracted through water droplets.

Moonbridge: A rounded bridge featured in Chinese meditation gardens. When their semi-circular shape is reflected in ponds and pools they resemble the moon.

Mooncalf: One who is carried away by an obsessive love of the moon. 2. A stupid, doltish person.

Moon Child: Preferred designation for some whose astrological sign is Cancer. Since the Moon rules Cancer, its traits are the same, but the moon doesn't have the negative disease connotation that cancer does.

Moondew: Said to be the first menstrual blood of girls gathered during a lunar eclipse and used to lay curses by Thessalian priestesses.

Moondog: Also known as a mock moon. The illusion of a second pale moon produced by the interaction of moonlight and moisture in the atmosphere.

Moondogger: Groupie of former Cleveland rock 'n' roll radio DJ Alan Freed, known as Moondog.

Moon-eye: A periodic inflammation of the eye suffered by horses.

Moon fish: Deep bodied, sharply compressed fish, usually silver or yellow, found in coastal areas of the Americas.

Moonflower: The ox-eye daisy.

Moon Grass: Wild tansy. Also called silverweed, argentine, goosewort, crampweed.

Moonie: Follower of South Korean religious leader Reverend Sun Myung Moon.

Moon Pillar: A rarely seen lunar halo, appearing like shafts of light above and below the moon, which is only visible when the moon is near to the horizon.

Moonraker: A sail carried above the sky-scraper of a ship. 2. A foolish person, specifically a native of Wiltshire, England after a story about two local men who saw the moon's reflection in the water and tried to pull it out with a rake thinking it was made of cheese.

Moonseed: Any of the Menispermaceae family, with twisting vines, small clusters of purple berries, and crescent-shaped seeds.

Moonshine: An illegal alcoholic beverage, presumably prepared and delivered by moonlight stealth under the cover of dark.

Moonshot: Launching a spacecraft to the moon.

Moonstone: A milky blue-green-whitish opalescent feldspar gem. Hindus believe that it is the physical manifestation of the rays of the moon. It is said to bring luck, cure epilepsy and nervousness, calm the nerves, cool the emotions and generally enhance female energies.

Moonstruck: Deranged in mind, crazy, wild and erratic.

Moontouched: Wandering mind, witless.

Moonwalking: A backward sliding dance step made popular by Michael Jackson.

Moonwort: A fern of the genus Botrychium.

Month: A period of time reckoned by the moon's revolution.

Moony: Dreamy, listless.

Numinous: Magical, enchanted.

Old man in new moon's arms: The faintly visible lunar disk cradled by the new crescent moon.

Selenium: Derived from Selene, an ancient name of the Greek moon goddess. Non-metallic element resembling sulfur and tellurium discovered in 1817; its electrical conductivity is increased by light, and is thus used in photographic telegraphy.

Selenocentric: Referring to the center of the moon; or to the moon as a center.

Selenodont: Having crescent-shaped molar teeth.

Selenograph: Chart depicting the moon's surface.

Selenography: The study and/or mapping of the surface topography of the moon.

Selenology: The study of lunar astronomy.

Senotropic: Turning toward the moon.

Selenotropism: Tendency to curve upward under moonlight.

Silly: From archaic Germanic meaning "blessed." Linked etymologically to Selene.

To ask for the moon: To crave for the unattainable. "I might as well wish for the moon as hope to get her." William Thackery, 1861

To know no more about it than the Man in the Moon: To know nothing at all.

To Moon: Northern European sport hunting which required the nighttime hunter to spot the animal in such a way that the moon shines directly behind it creating a halo effect — effectively outlining the creature for the shoot. 2. To show one's bare bottom for shock effect. Usually as part of a dare or other juvenile prank.

To moon about: To loiter, lounge in an idle, listless way.
"I did nothing whatever, except moon about the house and gardens."
JK Jerome, "Idle Thoughts," 1889

To moonlight: To hold down two jobs, including one at night.

To shoot the moon: To do a moonlight flit. To escape in the dark of the night. 2. A secret removal of furniture, etc. to prevent it from being repossessed.

Speaking of the Moon

The Moon by Any Other Name

African Pygmy	*Pe*
Ashanti	*Boshun*
Bahasa Indonesian	*Bulan*
Chinese	*Yuet*
Dutch	*Maan*
Danish	*Mane*
Egyptian	*Pooh*
Eskimo	*Tatkret*
Finnish	*Luna*
French	*la Lune*
German	*der Mond*
Greek	*Menos*
Hawaiian	*Mahina*
Irish	*Gealach, Luan*
Italian	*la Luna*

Japanese	*Otsukisama (reverent), Tsuki (common)*
Korean	*Tal*
Persian	*Mâh*
Peruvian	*Mama Quilla*
Polish	*Ksiezyc*
Polynesian	*Hina*
Portuguese	*a Lua*
Romanian	*Luna*
Russian	*Luna*
Sanskrit	*Gaus*
Serbo-Croatian	*Mjesec*
Spanish	*la Luna*
Swahili, Tutsi	*Mwezi*
Tatar	*Macha Alla*
Turkish	*Ay*
Welsh	*Ileuad*
Yiddish	*Levone*

Etceteras

Notes

(1) Lieber, Arnold and Sherin, Carolyn. "Homicides and the Lunar Cycle: Toward a Theory of Lunar Influence on Human Disturbance," American Journal of Psychiatry, 1972, 129 (1), 69-74.

(2) Jones, Paul and Jones, Susan. "Lunar Association with Suicide," Suicide and Life Threatening Behavior, 1970, 7 (1), 36.

(3) Surawicz, Frieda and Banta, Richard. "Lycanthropy Revisited," Canadian Psychiatric Association Journal, 1975, 20, 537-538.

(4) Andrews, Edson. "Moon Talk: The Cyclic Periodicity of Postoperative Hemorrage," Journal of the Florida Medical Association, 1960, 46 (11), 1366.

Permissions

Bibliography

Angel, Marie. *A Floral Calendar*. London: Pelham Books, 1985.

Aveni, Anthony F. *Empires of Time: Calendars, Clocks, and Cultures*. New York: Basic Books, 1989.

______ . *Conversing with the Planets: How Science and Myth Invented the Cosmos*. New York: Times Books/Random House, 1992.

Ayto, John. *Dictionary of Word Origins*. New York: Little, Brown & Co., 1990.

Belting, Natlia. *Calendar Moon*. New York: Holt, Rinehart & Winston, 1964.

Calvin, William. *How the Shaman Stole the Moon.* New York: Bantam Books, 1991.

Campbell, Joseph. *The Mask of God: Primitive Mythology*. New York: Penguin Books, 1959.

Coomaraswamy, Ananda K. and the Sister Nivedita (Noble, Margaret E.). *Myths of the Hindus and Buddhists.* New York: Dover Publications, 1967.

Geffen, Rela M., ed. *Celebration and Renewal: Rites of Passage in Judaism.* Philadelphia: Jewish Publication Society, 1993.

George, Demetra. *Mysteries of the Dark Moon*. New York: HarperCollins, 1992.

Guiley, Rosemary Ellen. *Moonscapes, A Celebration of Lunar Astronomy, Magic, Legend, and Lore.* New York: Prentice Hall Press, 1991.

Hall, Edward T. *The Dance of Life, the Other Dimension of Time.* Garden City, N.Y.: Anchor Press/Doubleday, 1983.

Harley, Timothy. *Moon Lore.* Rutland, Vermont & Tokyo, Japan: Charles E. Tuttle Co.: Publishers, 1970.

Harper, Howard V. *Days and Customs of All Faiths*. New York: Fleet Publishing, 1957.

Hughes, Paul. *The Months of the Year.* Ada, OK: Garrett Educational Corp., 1989.

Ickis, Marguerite. *The Book of Festivals and Holidays the World Over*. New York: Dodd, Mead & Co., 1970.

Jobes, Gertrude. *Dictionary of Mythology, Folklore and Symbols.* Vol. 1. New York: Scarecrow Press, 1962.

Katzeff, Paul. *Full Moons.* Secaucus, N.J.: Citadel Press, 1981.

Knappert, Jan. *The Aquarian Guide to African Mythology.* Northamptonshire, England: Aquarian Press, 1990.

Krupp, Edwin C. *Beyond the Blue Horizon.* New York: HarperCollins, 1991.

Kubler, George. *The Shape of Time.* New Haven and London: Yale University Press, 1962.

Landes, David S. *Revolution in Time.* Cambridge: The President and Fellows of Harvard College, 1983.

Leach, Maria and Fried, Jerome, eds. *Funk and Wagnalls Standard Dictionary of Folklore Mythology and Legend.* San Francisco: Funk & Wagnalls, 1949-50.

Long, Kim. *The Moon Book.* Boulder: Johnson Books, 1988.

Luce, Gay Gaer. *Biological Rhythms in Human and Animal Physiology*. New York: Dover Publications, 1971.

MacGregor, Geddes. *Dictionary of Religion and Philosophy*. New York: Paragon House, 1989.

Miller, John and Smith, Tim, eds. *The Moon Box.* San Francisco: Chronicle Books, 1995.

Monroe, Jean Guard and Williamson, Raymond A. *They Dance in the Sky: Native American Star Myths*. Boston: Houghton, 1987.

Opie, Iona and Tatem, Moira, eds: *Dictionary of Superstitions.* New York: Oxford University Press, 1992.

Passmore, Nancy F. W., et. al. *The Lunar Calendar: Dedicated to the Goddess in Her Many Guises*. Boston: Luna Press, annual since 1977.

Paungger, Johanna and Poppe, Thomas. *Moon Time: The Art of Harmony with Nature and Lunar Cycles,* translated by David Pendlebury. Great Britain: C. W. Daniel Company Limited, 1995.

Pegg, Bob. *Rites and Riots: Folk Customs of Britain and Europe.* Dorset, England: Blandford Press, 1981.

Rudaux, Lucien and De Vaucouleurs, G. *Larousse Encyclopedia of Astronomy.* New York: Prometheus Press, 1959.

Rush, Ann Kent. *Moon, Moon.* New York: Random House/Moon Books, 1976.

Sagan, Carl. *Cosmos.* New York: Random House, 1980.

Still, Henry. *Of Time, Tides and Inner Clocks.* New York: Pyramid Books, 1972.

Verdet, Jean-Pierre. *The Sky, Mystery, Magic and Myth*. New York: Harry N. Abrams, 1992.

Walker, Barbara G. *The Woman's Encyclopedia of Myths and Secrets.* San Francisco: HarperSan Francisco, 1983.

_______ . *The Women's Dictionary of Symbols and Sacred Objects.* San Francisco: Harper & Row, 1988.

Wong, C.S. *A Cycle of Chinese Festivities.* Singapore: Malaysia Publishing House, 1967.

Acknowledgments

My great gratitude goes to William Galison, the man who captured the faces and phases of the moon in a watch; who first envisioned this project and whose passion it has been for so long. To Margi Flanagan, Paula Franco, Naomi Grupp and Nancy Passmore for their eagle eyes, right hands and huge hearts. To Marina Bekkerman who stepped up to the plate with her creative genius. To Dalle Kaplan and the many many moons we have shared. To Shameike Thomas, the sun of my son and the moon of my heart. To the legions of moonstruck celebrants of every culture and every age. May we continue to congregate in the light and in the dark of the moon, and may we become enlightened and illuminated by the lovely lunar glow of that numinous celestial shape-shifter.

Photo: Trinh Thai

Donna Henes is an internationally recognized urban shaman, writer and artist whose joyful celebrations of celestial events have introduced ancient traditional rituals and contemporary ceremonies to millions of people in more than one hundred cities for thirty years. She is also the author of *The Queen of My Self: Women Stepping Into Sovereignty in Midlife, Celestially Auspicious Occasions™: Seasons, Cycles, and Celebrations,* and *Dressing Our Wounds In Warm Clothes,* as well as the CD, *Reverence To Her. Part 1: Mythology, the Matriarchy, & Me.* She publishes an acclaimed quarterly journal, *Always In Season: Living in Sync with the Cycles.* In addition to teaching and lecturing worldwide, Mama Donna, as she is affectionately called, maintains a ceremonial center, spirit shop, ritual practice and consultancy where she works with individuals, groups, institutions, municipalities and corporations to create meaningful ceremonies for every imaginable occasion.

For further information about Donna Henes' work, send for a calendar of upcoming events, a list of publications and services, and a complimentary copy of *Always in Season:*

Donna Henes
Mama Donna's Tea Garden & Healing Haven
PO Box 380403
Exotic Brooklyn, New York, NY 11238-0403
Phone/Fax: 718/857-2247
Email: MoonWatchBook@aol.com
www.DonnaHenes.net